Galvin

FOOD ENERGY
IN
TROPICAL ECOSYSTEMS

Food and Nutrition in History and Anthropology
Edited by
John R.K. Robson
Medical University of South Carolina
Charleston, South Carolina

Other volumes in preparation

ISSN 0275-5769

FOOD ENERGY IN TROPICAL ECOSYSTEMS

Edited by

DOROTHY J. CATTLE
University of California
Berkeley, California

KARL H. SCHWERIN
University of New Mexico
Albuquerque, New Mexico

GORDON AND BREACH SCIENCE PUBLISHERS
New York London Paris Montreux Tokyo

© 1985 by Gordon and Breach Science Publishers S.A. All rights
reserved. Published under license by OPA Ltd. for Gordon
and Breach Science Publishers S.A.

Gordon and Breach Science Publishers

P.O. Box 786
Cooper Station
New York, NY 10276
United States of America

P.O. Box 197
London WC2E 9PX
England

58, rue Lhomond
75005 Paris
France

P.O. Box 161
1820 Montreux 2
SWitzerland

14-9 Okubo 3-chome
Sinjuku-ku,
Tokyo 160
Japan

Library of Congress Cataloging in Publication Data
Main entry under title:

Food energy in tropical ecosystems.

(Food and nutrition in history and anthropology,
ISSN 0275-5769; v. 4)
    Bibliography: p.
    1. Diet—Tropics—Addresses, essays, lectures.
2. Nutrition surveys—Tropics—Addresses, essays,
lectures.   3. Food crops—Tropics—Addresses, essays,
lectures.   I. Cattle, Dorothy J.   II. Schwerin, Karl H.
III. Series.
TX360.T75F66   1985   613.2'0913   84-28998
ISBN: 2-88124-036-4

Cover photograph courtesy of Paul N. Patmore

TABLE OF CONTENTS

v

SERIES PREFACE

Two decades ago, the science of nutrition, as well as the various disciplines that contribute to nutritional knowledge, was fragmented into its many component parts and suffered greatly from a lack of interdisciplinary communication. Dominance by the "hard sciences" preserved a *status quo,* and the interests and ongoing research at that time often did not reflect the concerns of persons working in the Third World. A prevailing trait, and one that continues to some degree to this day, is the "remification phenomenon" characterized by research that continually expands and branches out from a main theme.

The situation can be likened to studies of a tree in which the scientist generating new knowledge relating to a single leaf becomes so preoccupied with his specialty that he loses touch not only with the significance of the twig to which the leaf is attached, but also with the significance of the limbs and even the trunk itself. In order to effectively use the detailed knowledge collected in research, there is a need to understand not only the significance of the tree but also how the leaf, twig, bough, trunk, tree, copse, or forest fits into the total environment. There is reassuring evidence that science is now accepting this concept and is recognizing the need to integrate information from a wide variety of sources and to understand the implications. Acceptance has been a slow process, but it is gratifying that nutrition research based on malnutrition in the Third World countries has served as a major catalyst.

After Cecily Williams first described kwashiorkor in the early 1930s and attributed this to protein depletion, the etiological approach prevailed. This involved consideration of the immediate cause, and, on this basis, kwashiorkor was deemed to be due to inadequate weaning. It was argued, therefore, that the cure was obviously the provision of protein as a weaning supplement. This conclusion, which proved to be erroneous, was never challenged and, as a consequence, enormous sums of money were spent providing weaning foods to the infants and children of the Third World. Although many of these efforts were ineffective, limited success was achieved, and some of the

vii

poorly nourished were restored to health. Unfortunately, after being cured they were discharged from the hospital to the same environment that caused the problem. It took years to realize the futility of this, but, eventually, attention turned to the remote causes of the food deficits.

In this "ecological approach," attention was focussed on the circumstances leading to the shortage of protein. Once the significance of this concept was realized, it became obvious that the background of and the education and training in the dominant nutritional sciences were often inappropriate for collecting the information needed. Infants were found to be protein depleted because prevailing tribal law and custom had not adapted to modern life, or because of trade tarriffs, or inappropriate agricultural policies. Interest in these problems spread to behavioral scientists, economists, and agronomists; however, opportunities for these other disciplines to publish their observations and air their views were lacking,

The journal *Ecology of Food and Nutrition* partly met this need, but a more eclectic approach was indicated. In an effort to understand the total picture of nutritional problems, the present series was begun. Earlier volumns in this series examined the ecological (i.e., the remote) causes of famine, of food ecology and culture, and infant feeding in the South Pacific. In these publications anthropologists have assumed an important role by providing a better understanding of food behavior. The operational horizons of nutrition have extended even further, however, and now anthropologists and geographers are demonstrating the nutritional nuances of agricultural systems in different geographical locations. *Food Energy in Tropical Ecosystems* is another major step toward providing information on the various influences that affect the food supplies of populations, population groups, and individuals. It should do much to foster better communication between scientists involved in nutrition, it should promote a better understanding of the complexity of the origins of current nutrition problems throughout the world, and it should help individual researchers see how their work fits into the total picture.

John R.K. Robson
Charleston, South Carolina

PREFACE

Contributions to this volume address the question: What is the importance of food energy for human well-being in tropical ecosystems? This question emerged from initial discussions in which both editors evaluated possible subjects for papers based on their long-standing interest in nutrition. From the very beginning it was evident that each editor was identifying issues and areas of inquiry from related, but different, perspectives. The basic question raised above seemed to adequately cover these differing viewpoints.

Research by Karl H. Schwerin has emphasized Karinya subsistence practices in the tropical lowlands (Denevan and Schwerin 1978). From related interests in the development of native cultures in the American tropics has come a realization that understanding human adaptation in one tropical region necessitates understanding the processes of adaptation to tropical conditions in other parts of the world as well. Consequently, his research continues to emphasize tropical resources and the ways in which these are used by the human population.

Dorothy J. Cattle has been more directly concerned with nutritional questions, making this a principal focus of her fieldwork on changing nutritional patterns among the Miskito Indians of coastal Central America (Cattle 1977). Her research in nutritional anthropology has examined the implications of dietary diversity for nutritional security (Cattle 1976a) and has explored conceptual-theoretical frameworks for nutritional anthropology (Cattle 1976b). Given the commitment to a biocultural framework by nutritional anthropologists, Cattle also has interests in human-environmental interactions (Cattle et al 1981). At the time this book went to press, she was in Kenya investigating the relationships between levels of food intake and physiological function, as well as social behavior. Beginning with the sociocultural dynamics of resource situations, her teaching and research have proceeded analytically and descriptively to resource use and management issues.

Therefore, our experiences and research interests naturally led us to consider the following papers in terms of three major areas: climatology, nutrition, and natural resources.

This volume originated with a symposium at the annual meeting of
the American Anthropological Association. In recent years, this
meeting has been highlighted by several symposia on the American
tropics, and similar sessions have been presented on other major
tropical regions. Concentration on regional analysis has proved both
stimulating and productive. We believed that these discussions had
progressed to the point where valuable insights might emerge through
bringing together people working in different geographical areas, so
that we might better perceive those aspects of human behavior and
adaptation that are common to the tropics. Because of our own
mutual interests in food resources and nutrition, we decided to make
this the theme of a single, well-integrated symposium presented at the
79th Annual Meeting of the American Anthropological Association,
held December 3–7, 1980, in Washington, D.C.

As organizers, we made a concerted effort to balance the
presentation on the basis of several dissimilar criteria. In the first
place, we sought to include a balance of papers dealing with each of
the four major tropical regions: the Americas, Asia, Africa, and the
Pacific. We also wanted to encourage a fuller dialogue among ecologically oriented anthropologists and those investigators who are
more specifically involved with nutritional questions. Finally, we were
concerned to ensure status parity, and thus sought to include an
equal number of women and men, junior and senior scholars, and
participants drawn from different regions and institutions around the
United States. We were not wholly successful in satisfying all these
criteria but feel that these extra efforts and attention to equity are
warranted within contemporary academe and are inherently worth-
while.

The quality of the papers met all preliminary reviews. More
importantly, the individual papers related to each other in a way that
reveals some important things about subsistence systems, nutrition,
and human adaptation in the tropics. We, the symposium organizers,
felt that, both individually and collectively, these contributions could
be expanded into an edited volume. Research on the symposium topics
will continue to grow significantly in the future, and such a volume
would be of interest to both professionals and students in cultural and
biological anthropology, cultural ecology, nutrition, tropical studies,
and related research fields. The volume could serve as a source for
detailed case materials and as a guide for continuing research.

All of the symposium participants agreed to expand their oral
presentations to a form appropriate to these purposes, namely, a

fuller treatment of their material and an expanded bibliography so that the volume would be bibliographically useful as a research guide. We solicited additional papers in order to cover salient topics. The result is an instructive and useful set of contributions dealing with the four major tropical areas and with a range of related research problems that extends from specific questions surrounding a single dietary staple, through an emphasis on various nutritional considerations, to the broad issues of adaptation and change in the tropics. In addition to this topical structure, we have attempted in several other ways to tie these contributions together and to enhance the general usefulness of the book as an information source and research guide.

The Foreword contributed by Angela C. Little considers several nutritional questions related to the tropics and discusses these with respect to the nutritional issues raised throughout this book. Research directions are dealt with from a food science and nutritional science point of view.

The papers, provide empirical case materials drawn from the four major tropical regions. We have grouped them according to general perspectives. The first set is concerned with specific questions relating to a single dietary staple. The study by Dufour deals with the harvesting, processing, and cooking of manioc among the Tatuyo in the Amazonian lowlands of Colombia. She shows not only that manioc is the major source of energy in their diet, but also that it takes up a major portion of the time and energy devoted to daily activities. Kunstadter compares techniques of swidden and irrigated rice farming in northwestern Thailand and discusses the factors involved in achieving a satisfactory level of production. He shows how, in recent years, weakening village-level authority, population pressure, and better economic conditions have encouraged the expansion of irrigated agriculture at the expense of swidden farming.

The second set of papers deals with the production of several subsistence items. In her comparison of two Indian groups of north-central Peru and the Peruvian immigrants recently settled in the area, Berlin shows how cultural differences influence the diets and agricultural strategies of all three. The Indians tend to rely on root crops and fishing, whereas the colonists are more reliant on grains and domestic animals. She predicts that population growth and external pressures for cultural change will eventually push the Indians to develop a subsistence strategy more like that of the colonists. Wilson evaluates Malay rice agriculture in a fishing village on the east coast of Malaysia. While rice provides the bulk of their calories, it is complemented by the addition of the fish that is usually consumed with it. These are supplemented by many other carbohydrate foods. The result is a population that seems to be adequately nourished, although the increased consumption of refined carbohydrates since World War II provides less of the needed nutrients than did the traditional foods. Anderson compares a pair of subsistence systems on the west coast of Malaysia. Again, differences in Chinese and Malaysian food preferences and diets reflect cultural differences. In their dependence on rice agriculture complemented by fish, the Malays are like those described by Wilson. However, Anderson discusses in greater detail the diversity of fruits, spices, and other crops that supplement the diet. In contrast, the Chinese grow vegetables both for subsistence use and for commercial sale. There is less use of spices and more emphasis on livestock, especially pigs. Both systems seem well adapted to local conditions, but optimum production could probably be achieved through an exchange of crops and farming techniques between the two ethnic groups. This is unlikely, however, due to ethnic tensions between the two groups and official Malaysian policies that encourage different spheres of economic activity and public involvement for each one.

Dewey's paper deals with another agricultural system characterized by a great deal of crop diversity: that of southeastern Mexico. A large-scale agricultural development project in the 1960s has encouraged a shift to cash crops and cattle pasturage, with a concomitant decrease in crop diversity and self-sufficiency and an increase in dependence on purchased foods. She demonstrates that these developments have produced a corresponding decrease in dietary diversity, dietary quality, and nutritional status, as measured among preschool children. To prevent such repercussions in the future, she urges the elaboration of a development philosophy that will stimulate more progressive forms of agricultural change.

The final group of papers deals with more general questions of adaptation to tropical subsistence and resource ecosystems and especially with the impact of technological and economic change on those systems. Each also deals with a different staple crop. The Fleurets provide an ethnohistorical and ethnographic study of the place of the banana in the culture and diet of the Shambaa of northeastern Tanzania. In precolonial times it was the major staple crop. Introduced grains and root crops have significantly reduced the importance of the banana as a subsistence crop, but it continues as a secondary domestic foodstuff and is still important in local trade.

McCutcheon describes the intensive techniques of cultivating taro on the Palau Islands of Micronesia from the time of *initial* contact, when the density of the population was high, through a period between 1900 and about 1960, when the population density was low. Now, as the population has begun to grow again, taro agriculture has shifted from intensive swamp cultivation to swidden techniques and has been replaced by other, less intensive crops. This apparent paradox in the curve of agricultural intensification is explained in terms of changes in the social, economic, and dietary role of taro and the relationship of women laborers to agricultural production. Throughout the Pacific, social and agricultural change has been associated with the widespread adoption of manioc (cassava) as a major subsistence crop. The introduction of manioc is often viewed negatively, but Thaman and Thomas take a much more positive view. In their detailed assessment of the crop, they consider both positive and negative qualities as these relate to such factors as economics, labor requirements, environmental tolerance, productivity and nutritional characteristics. They conclude that manioc could contribute strongly to nutritional improvement by making local populations less dependent on imported foodstuffs, thereby facilitating greater economic independence as well.

The book concludes with an Afterword that comments on the empirical case studies and complements the Foreword. This piece, written by Antoinette Brown, a nutritional anthropologist with field experience in Mexico, serves as a bridge between the two fields of nutrition and anthropology. It also comments on the relation between nutrition and development.

Tropical research needs to provide some means to facilitate understanding of particular aspects of the tropics and its food resources for those more familiar with temperate zone adaptations. Through the photographs and maps some information may be gained on tropical foodstuffs, production, and environment, as well as resource distribution. As a further aid, we have included a final chapter that details tropical foodstuffs and, most usefully, provides information on the botanical characteristics, origins, agricultural requirements, nutritional composition and bibliographic sources.

A major link between anthropology and nutritional sciences is resources. Anthropology, a discipline well known for its eclecticism, has traditionally included aspects pertinent to resources. Traditionally nutritional science has been more narrowly focused on food items as resources, but becomes broader in its treatment of resources when examining agricultural and public health questions. Both disciplines now recognize the necessity for consistently considering biophysical and social factors in the interplay between a human population and its environments in terms of resources. The multiple interactions between a human population and its environments are likely to include repetitious processes and cycles as well as various types of associations, responses, and changes among all components of the tropical ecosystem. Thus, circularity is a more apt characterization than most kinds of linear sequences. The biological and social integrity of such systems are of interest to nutritional and anthropological investigators concerned with practical solutions to situations not yet well understood. There are many ways to proceed under these circumstances. We think that it is initially useful to confront the following areas: complexity of food resource phenomena; enhanced communication within and among the various fields and specialties involved in tropical resource questions; and general research considerations regarding data, methodology, and theory. Researchers in nutrition and anthropology need to provide broadly applicable concepts and common points of reference and to indicate the potential for further elaboration and interpretation of their work. More extensive evaluations of material such as that presented in this volume are expected to result in modifications intended to help us better understand or more effectively approach tropical food resource issues.

REFERENCES

Cattle, Dorothy J. 1976a. Dietary Diversity and Nutritional Security in a Coastal Miskito Indian Village, Eastern Nicaragua. pp. 117–130 in *Frontier Adaptations in Lower Central America*, ed. by Mary W. Helms and Franklin O. Loveland. Philadelphia: Institute for the Study of Human Issues

———— 1976b. An Alternative to Nutritional Particularism. pp. 35–45 in *Nutrition and Anthropology in Action,* ed. by Thomas K. Fitzgerald. Assen: Van Gorcum

———— 1977. *Nutritional Security and the Strategy of Purchasing: The Coastal Miskito Indians, Eastern Nicaragua.* Ph.D. dissertation, University of New Mexico, Albuquerque.

Cattle, Dorothy J., Charles H. Carroll and David E. Stuart. 1981. *Ethnoarcheological Investigations of Sheepherding at the Pueblo of Laguna.* Environmental Quality Services, Albuquerque Area Office. Albuquerque, New Mexico: Bureau of Indian Affairs

Denevan, William M. and Karl H. Schwerin. 1978. Adaptive Strategies in Karinya Subsistence, Venezuelan Llanos. Antropológica 50:3–91.

INTRODUCTION

Whenever we discuss questions having to do with food and agriculture at any level and with whatever emphasis, we are, in fact, discussing nutrition and nutritional status. Conversely, whenever we discuss questions having to do with nutritional requirements and status, we must acknowledge, if only by implication, the nutritional precursor, food: its production, processing, and utilization. Thus, we can state with conviction that all the papers in this collection, no matter what their primary focus or general perspective, are in fact papers pertaining to human nutrition.

It is not my intention to identify and discuss the salient nutritional questions raised by the individual contributors—that would be redundant. Instead, I would like to pursue some of the thoughts that occurred to me as I read the individual contributions comprising this book. Reference to certain papers will be made from time to time, only because those papers triggered a particular thought. They are singled out for no other reason. All of the papers together are responsible for the general tone of these ruminations.

As I read the papers, I found myself thinking more and more about the people discussed by the various authors: their daily activities, division of labor, the climatic and environmental variables they must contend with, their general health, and so on. And it became very clear that I was reading about men and women not representative of the reference adults defined by the Food and Nutrition Board (FNB) of the U.S. National Academy of Sciences, National Research Council (NAS-NRC) [a man of 70 kg (154 1b), 175 cm (5′10″) and a woman of 58 kg (128 1b) 160 cm (5′4″)] with the recommended energy intakes predicated on size, a mean ambient temperature of 20°C (68°F), clothing suitable to temperature, and activity levels corresponding to light occupation. For such reference adults, the daily calorie requirements are estimated at 2,700 Cal (11,300 kJ) for men and 2,000 Cal (8,400 kJ) for women, between the ages of 23 and 50 years.

Table of recommended daily allowances (RDA) for energy and essential nutrients are based on populations in our Western, industrialized, climatically temperate (or temperature-controlled) life patterns throughout the life cycle, with the characteristic growth and development patterns of infants and children comprising the cultural norm.

The first question I ask is: Do these concepts and recommendations pertain to nonindustrialized populations living in tropical ecosystems?

With regard to differences in size and energy requirements between males and females, I question whether these differences are as apparent among many of the tropical peoples as among Northern and Western industrialized peoples. Observations and intuition suggest that the size differential between the sexes is more prominent among populations where differences in labor and activity levels are culturally determined, where masculine strength and, therefore, size is correlated with masculine dominance and feminine delicacy and diminutive size are correlated with feminine submission and dependency. In such cultures, where the male has been traditionally viewed as the hunter, the warrior, the fighter, evolutionary selection would favor larger, heavier males and smaller, lighter females. In the tropics, where subsistence agriculture has minimized sex-differentiated activities, men and women do not seem to differ so markedly in body size and muscular development. If this is indeed the case, the energy and nutrient RDAs for reference men and women may lose their significance. To answer my first question, then, it will be necessary to determine the extent to which nutritional requirements are modified, based not on a sexual size and activity differential, but on less sexually differentiated populations where ecological factors and labor-intensive activities may play more prominent roles.

The second question has to do with how one views the requirements of a nutritionally adequate diet for tropical populations. Can we apply the traditional approach to ensuring a well-balanced diet based on daily selection from several food groups (the number varying over time and recently tending to expand once more from four to six)?

We learn—and teach—that foods should be selected from diverse groups such as fruits and vegetables; cereals, breads, and grains; meats, poultry, eggs, and fish; dry peas and beans, such as soybeans, kidney beans, lima beans, and black-eyed peas; and milk, cheese, and yogurt. However, on reading the papers in this collection, one is struck by the paucity of variety with which the various tropical populations sustain themselves. Typically, carbohydrate-rich staples —rice, corn, cassava, taro, yams, and bananas—constitute "cultural superfoods" and provide the dominant share of caloric intake. In fact, cassava (manioc) is gaining in importance as the major source of dietary calories for increasing numbers. The presence of cyanogenic glucosides in cassava, particularly the so-called "bitter" variety, necessitates careful processing to remove the toxic constituent. The potential impact of long-term ingestion of large quantities of food containing even vestigial traces of cyanide has yet to be elucidated. Other questions regarding the effect of cassava ingestion—having to do with the effect of cyanide-containing foods on the alleviation of sickle-cell anemia symptoms and with the goiterogenic properties of cassava exacerbated by the iodine deficiency in some areas —also remain to be answered.

Other foods are eaten as supplements to the main carbohydrate-rich staple, leading to a food intake pattern that provides marginal nutritional intake and dietary monotony. As Brown points out in her paper on tropical nutrition, dietary inadequacy and monotony are not confined to tropical ecosystems but are characteristic of poverty and underdevelopment in general. Although many nutritional text and reference books and far more primary literature sources deal with questions of nutritional status in "third world" populations, the tendency is to think of meeting nutritional requirements in terms of foods familiar to us. Easily accessible food composition tables list hundreds of foods, but do not acknowledge the existence of cassava, breadfruit, yams (not to be confused with sweet potatoes!), palm oil (a rich source of Vitamin A), winged beans, the many varieties of bananas (discussed below), the edible green leaves from taro and manioc (cassava), or such "exotic" fruits as cashew fruit and guava (important sources of ascorbic acid) and mango (rich in carotene). Bananas can be found in any food composition table. They are, after all, a popular item in the Western diet. Yet only one variety of banana predominates in our market place; only occasionally do we find plantains or red bananas in the local supermarket. Yet it is clear from Fleuret and Fleuret's discussion that the representative analysis of Tanganyika (now Tanzania) bananas differs in some important respects from the analytical data presented in the FAO-PHS Food Composition Tables for use in Africa, 1968, and from analyses presented in American textbook food composition tables. The major important differences lie in carotene and ascorbic acid content. Differences in the former suggest that 1 kg (2.2 lb) of Tanganyikan bananas would meet the daily Vitamin A requirement, in contrast to 4 kg (8.8 lb) of the nonspecific banana in the food composition tables, while differences in the latter suggest that the Tanganyikan banana is more than 1.5 times as effective in meeting the ascorbic acid requirement, with 300 g (10.5 oz) of the one and 500 g (1 lb 2 oz) of the other being comparably effective.

Varietal differences can, therefore, play an important role in determining nutritional adequacy. It seems reasonable to surmise that nutritional adequacy might be more easily approached if diversity of food intake is encouraged based on the previously known or newly determined composition and properties of indigenous foods. Without prominent disclosures and easy accessibility of such data, nutrition planners cannot be too severely criticized for stressing and advocating western-style processed foods. This point is well taken by McCutcheon in her paper on agricultural change in Palau, in which she expresses concern about the diets of elementary school children because the federally funded lunch program relies heavily on canned fish, canned "fricasee of chicken wings," and rice. She suggests that native produce, specifically fresh fish, taro, and cassava, "should be offered instead both to upgrade the nutritional status of the lunches and to encourage productivity of these foods on a local level."

Dewey, in her paper on the nutritional consequences of the transformation from subsistence to commercial agriculture (Tabasco, Mexico), makes a similar point when she argues that this transformation resulted in a marked reduction in the production of subsistence crops by rural families, a decrease in crop diversity and a concomitant increase in dependence on outside sources of food. These shifts resulted in a decrease in dietary quality and diversity and the deterioration of nutritional status, particularly among small children, where a major negative effect was related to increased sugar consumption. When dietary diversity was superior due to the rural families' access to land and, therefore, their ability to grow crops for home consumption, an increase in dietary quality was attained. Thaman and Thomas, in their paper on cassava in the Pacific Islands, also identify "modernization" with drastic changes in traditional food systems and suggest that cassava offers an important alternative to increasing dependence on foreign foods in the Pacific, if, at the same time, the traditional agricultural and nutritional diversity characteristic of the Pacific Islands is preserved and their ecological integrity protected.

It is clear from these and other studies that increased understanding of indigenous foods and knowledge of their composition and properties are essential to the promotion of nutritional well-being among Third World peoples.

The third question flows directly from the first two. With a different array of foods to choose from, and particularly with the prominent use of high-fiber foods, how is the bioavailability of essential nutrients affected?

This is an important question since the limits to dietary diversity do not allow, as a rule, conformity with the generous FNB-RDAs and their wide margin of safety.

The amount of calcium required to ensure good bone mineralization during development and to prevent osteoporosis and osteomalacia remains controversial, and, even in our culture, where food sources rich in calcium are readily available, problems regarding bone integrity loom large. How does lower protein intake and/or the intake of protein of plant rather than animal origin affect calcium intake—that is, does it lower the requirement? What about increased phytate and fiber intake? What about vitamin D, which is hard to get unless the diet is supplemented? What about exposure to sunshine? Questions regarding the bioavailability of iron and trace minerals can also be asked, since cations are liable to be sequestered or chelated by phytate and dietary components present in fiber, and thus rendered less available.

The fourth question is really a series of questions and incorporates cultural and ecological factors. How many different foods are normally available and/or eaten? How do food beliefs affect ideas of appropriateness? For whom? Under what conditions (pregnancy, lactation, old age, illness)? Are there food prohibitions due to taboos or religious constraints? Do cultural practices determine the order of food offerings within the family? For example, do the father and other males have first choice, with boys second, girls third, and the mother and other adult women last? Are children fed separately from adults and therefore prevented from receiving essential foods? With respect to this issue, Thaman and Thomas attribute nutritional deficiencies to the practice of putting children to bed with a supper composed mainly of cassava before the adults have their evening meal, thus denying children access to important sources of nutrients. In other instances they cite the simple inability of a small child to dip its wad of cassava paste into the stew pot as a factor in nutritional deprivation.

With regard to the question of adequate protein intake, both quantitatively and qualitatively, have all possible local sources been seriously considered? Is there any way that grubs and insects can be incorporated into diets in a palatable, aesthetically acceptable fashion? What is known of their nutritional potential suggests that these invertebrates are as capable of supplying high-quality protein as other essential nutrients to the diet.

The sixth set of questions has to do with the role of locally fermented beverages as a somewhat underplayed source of certain essential nutrients (not covered in this book). To what extent can the additional protein and vitamins contributed by the fermentation organisms improve the marginal dietary intake? How counter-productive is the substitution of soft drinks and clarified processed beverages for these home brews? Do the nutritional advantages outweigh the disadvantages related to alcohol intake?

The final question, having to do with nutritional status in tropical ecosystems, concerns the prevalence of intestinal parasites. How does a population of hungry animals in the gastrointestinal tract affect nutrient requirements? I pose the question in this way because I think that it is more important, or perhaps more realistic, to consider the host-parasite relationship as the given than to suggest that the parasite be eradicated to protect the host. Reinfection is all too easy.

The seven categories of nutrition-related questions presented here are not exhaustive; they merely represent questions generated in my mind from exposure to the information presented in this book. I hope that at least some of these questions will provide an impetus for further or new research designed to improve the lot of a large proportion of the world's population.

Much needed research in food science and technology remains to be done. Though not an idea original with me, I wish to stress the importance of developing appropriate technologies both to improve agricultural practices, which, as McCutcheon points out *vis-a-vis* taro, may not necessarily be the most 'efficient' in terms of yield/area, and to provide better means of processing foods for consumption consistent with the constraints imposed by limited energy supplies. Here I would like to suggest again the importance of promoting microbial conversion of raw materials to more stable, fermented products, whether milk or fruit or vegetable beverages and products. Little energy is required for these conversions, and nutrient value is often increased.

The major hurdle that remains, once observations and mea-surements are made, once processes are modified or developed, is doubtless the most difficult of all: translating observations and experimentation to action. The importance of alleviating under-nutrition is unarguable; the importance of finding the means to ex-pedite the alleviation of undernutrition is equally unarguable. While the means remain elusive, certain hopeful signs are provided by the increasing cooperation among nutritionists, food scientists, anthropologists, health scientists, and policy and planning specialists. The process will not be easy; the stakes, however, are high.

Angela C. Little
University of California, Berkeley

Contributors

Eugene N. Anderson, Jr., Ph.D.,
Department of Anthropology,
University of California,
Riverside, California 92521, U.S.A.

Elois Ann Berlin, Ph.D.,
Dept. of Family, Community and Preventive Medicine,
Stanford University,
Stanford, California 94305, U.S.A.

Antoinette Brown, Ph.D.,
Dept. of Community Health Nutrition,
Georgia State University,
Atlanta, Georgia 30303, U.S.A.

Kathryn G. Dewey, Ph.D.,
Department of Nutrition,
University of California,
Davis, California 95616, U.S.A.

Darna L. Dufour, Ph.D.,
Department of Anthropology,
University of Colorado,
Boulder, Colorado 80309, U.S.A.

Anne K. Fleuret, Ph.D.,
Catholic University,
Rockville, Maryland 20833, U.S.A.

Patrick Fleuret, Ph.D.,
Bureau for Policy and Program Coordination,
U.S.A.I.D.,
Washington, D.C., U.S.A.

Peter Kunstadter, Ph.D.,
East-West Population Institute,
East-West Center,
Honolulu, Hawaii 96848, U.S.A.

Angela C. Little, Ph.D.,
Department of Nutritional Sciences,
University of California,
Berkeley, California 94720, U.S.A.

Mary McCutcheon, Ph.D.,
Dept. of Anthropology/Geography,
University of Guam Station,
Mangilao, Guam 96910

Karl H. Schwerin, Ph.D.,
Department of Anthropology,
University of New Mexico,
Albuquerque, New Mexico 87131, U.S.A.

Randolph R. Thaman, Ph.D.,
School of Social and Economic Development,
University of the South Pacific,
Suva, Fiji

Pamela M. Thomas,
Department of Geography,
Australian National University,
Canberra City, ACT,
Australia

Christine S. Wilson, Ph.D.,
Dept. of Epidemiology and International Health,
University of California,
San Francisco, California 94143, U.S.A.

PART I
DIETARY STAPLES

CHAPTER 1

MANIOC AS A DIETARY STAPLE: IMPLICATIONS FOR THE BUDGETING OF TIME AND ENERGY IN THE NORTHWEST AMAZON[1]

DARNA L. DUFOUR
Department of Anthropology
University of Colorader, Boulder 80309
and
Fulbright Lecturer
Departamento de Biologia
Universidad Pedagógica Nacional
Bogotá, Colombia

ABSTRACT

Manioc (*Manihot esculenta* Crantz) is the staple crop and the primary source of food energy in the diet of Tukanoan Indians in the Northwest Amazon. It provides over 70% of the dietary energy. This paper analyzes the time and energy expended in the harvesting and processing of manioc and its preparation into culturally acceptable forms of food. The work involved is allocated to women members of household groups. The time budgets of adult women were assessed through 24-hour time motion studies and the metabolic energy expended assessed through indirect calorimetry. The results indicate that the harvesting, processing and preparation of manioc foods account for approximately 26% of the total daily time budget and 45% of the total daily energy budget. Because manioc provides such a large proportion of the food energy in the diet, these relatively high time and energy inputs are critical to the maintenance of adequate food energy in the household

1

INTRODUCTION

This paper deals with the implications of the use of bitter manioc[2] (*Manihot esculenta*) as a principal source of dietary energy. In particular it examines the time and energy expenditure involved in the cultural context. It is attempted here to demonstrate that in this economy, where manioc is the single most important source of food energy, and where female members of domestic groups assume the responsibility for manioc production, the greater part of a woman's time and energy is allocated to cultivation and processing. Further, the consistency of time and energy inputs by women is an important factor in the maintenance of household self-sufficiency.

Fieldwork among the Tatuyo was carried out in the Colombian Vaupés between November 1976 and April 1978. (See Fig. 1.) The Tatuyo are one of the various exogamous, partilineal groups of Tukanoans in this region. Like other such groups they are rootcrop horticulturalists relying on manioc as a caloric staple. Dietary animal protein is obtained from fish and to a lesser extent from game and invertebrates. A variety of wild vegetable products are collected and also make important contributions to the diet.

Indigenous settlements in the Vaupés include both traditional longhouses and multi-house villages. The data reported here were gathered in Yapú, a multi-house village in the headwaters of the Papurí River. In 1977 the village had a population of about 140 persons, divided into some 25 household groups. It was one of the largest settlements in the upper Papurí region. In this paper the inhabitants of Yapú are referred to as Tatuyo, although some speakers of other Tukanoan languages were also resident in the village.

The Vaupés is a transitional zone of humid to very humid tropical forest. The terrain is gently undulating, with the forest cover broken by patches of scrub woodland or *caatinga*. Mean annual temperature is relatively high at about 26°C, and rainfall abundant at 3500 mm/yr. Although seasonal differences in temperature and rainfall are not well marked, there is a dry season of slightly less rainfall from November to February.

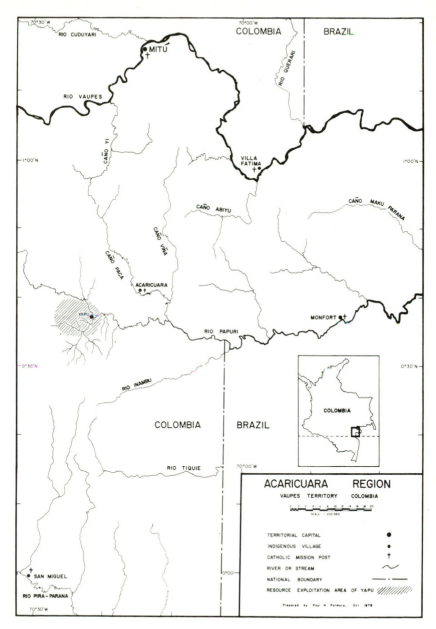

Figure 1. Map of Colombian Vaupés showing location of Yapú fieldsite. (Dufour, 1982).

METHODS

As part of an energy flow study, detailed short term case studies of four households were undertaken in the month of November. Each case study included estimates of total household food consumption for a four day period and time and energy expenditure of adult female heads of household during three days. The methods used are discussed in detail in Dufour (1981) and are summarized briefly below. Household food consumption was estimated using a weighed larder method in which a continuous record was kept for four days of all food inputs into the household and all food outputs. Household food energy consumption was taken as the difference between food inputs and outputs after controlling for the food stored in household larders.

The estimation of time expenditure was done through 3-day time-motion studies. Rates of energy expenditure in important subsistence activities were determined through indirect calorimetry on a sample of 10 village women. In this method the calories used in performing a task are calculated from the volume of oxygen consumed. The usefulness of this technique has been demostrated by a number of recent field studies. (See, for example, Norgan, Ferro-Luzzi and Durnin 1974; Montgomery and Johnson 1977.) The energy expenditure values were then used in conjuction with the time-motion studies to calculate total daily energy expenditure, and energy expenditure in specific activities. Rates of energy expenditure for activities not measured were taken from the literature.

THE ORGANIZATION OF SUBSISTENCE

In villages such as Yapú, the household is the basic unit of food production and consumption as well as social reproduction. It is this group which is expected to be self-sufficient in terms of nutritional and material needs.

The composition of households is variable. However, they always include one adult man and one adult woman, usually a married couple. Whether the household is a single nuclear family or a more complex group, it is characterized by a division of labor in which the woman, or women, harvest and prepare the carbohydrate staple of the diet, manioc. The man, or men, assume the responsibility for the procuring of animal protein and the manufacture of most household goods. The interdependence created by this division of labor between women and men integrates them into a productive unit. Men fell and burn gardens but do not (except under very unusual circumstances) harvest or prepare manioc. Women harvest and prepare manioc into a variety of foods using a very specialized basketry which only men make. Both men and women gather wild vegetable products and small invertebrates, but only men hunt and kill the larger fish and game that are complementary to the vegetable foods produced by women.

As a consequence of this division of labor, it is the women who produce the greater portion of the calories. This energy is derived principally from manioc. As a source of food energy, manioc is relatively abundant in relation to need, shows little seasonal variation in availability and is distributed in the environment in a known pattern. It is a valuable source of carbohydrate but contains little vegetable protein and virtually no fat (Jones 1959:6). In addition, raw manioc tubers are toxic to varying degrees, due to the presence of cyanogenic glucosides (Coursey 1973:28–30).

The subsistence activities of men are focused on local fish and animal populations. The energy available in these food chains is relatively less abundant and more dispersed in the environment, but it is particularly important because of its nutrient density and high protein quality. The behavior of animal and fish populations produces seasonal and temporal variations in availability. These variations, coupled with the somewhat irregular subsistence activities of men, lead to considerable day-to-day fluctuations in energy and nutrient flow through the human population from these sources.

The wild vegetable products and small invertebrates collected by both men and women, are dispersed in the environment and are seasonally important as energy sources. In contrast to cultigens, these products are characterized by high caloric density and relatively small edible portions.

SOURCES OF DIETARY ENERGY

Figure 2 shows the results of the household energy consumption studies. These values represent the food energy consumption of four households over a four-day period, expressed as a percentage of the total intake. Manioc accounted for 83.3% of the total caloric intake. Fish was the second most important source of calories, providing about 7.8% of the intake. Fish, meat and insects also provided animal protein, but all of these sources taken together accounted for only

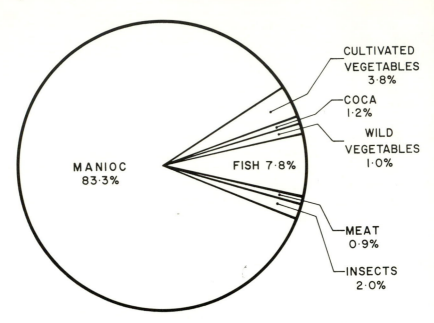

Figure 2. Mean values for energy consumption in four Yapú households measured over
a four-day expressed as a percentage of total energy consumption.

10.7% of the total caloric intake. Cultivated vegetable products, including fruits and tubers (except manioc) accounted for merely 3.8%. Wild vegetable products[3] were of minor importance at this time of the year and constitute only another 1.0%. Coca,[4] which is used almost exclusively by men and is here considered both a food and a drug, accounted for about 1.2% of the total consumption.

MANIOC AND MANIOC FOODS

That manioc is of extraordinary importance as a source of calories is clear. The production of this food energy, and the consistency of this energy source in the diet depends on the day-to-day work of women. This work is structured by the characteristics of manioc and the way in which it is used in the Vaupés. Three factors are particularly important: first, manioc requires extensive processing and cooking, both to increase palatability and decrease toxicity. Second, manioc is stored unharvested in the ground. No more than a few days' supply of

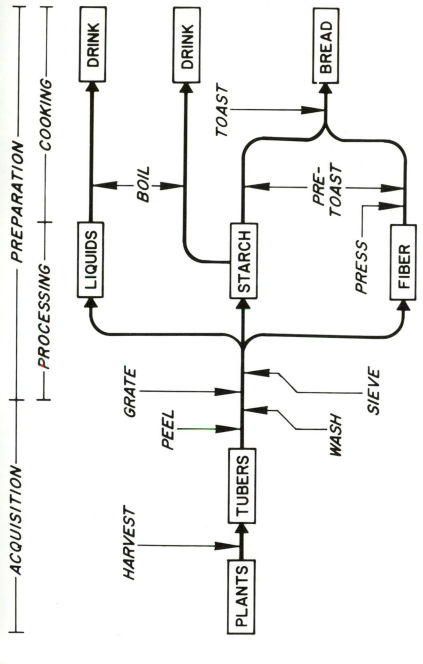

Figure 3. Schematic diagram of the acquisition, processing and preparation of manioc as routinely done in the Colombian Vaupés.

processed manioc products are held in household larders, and most cooked products are prepared fresh daily. Finally, in contrast to many other areas, both harvesting and processing are carried out on the same day.

In the Vaupés manioc is elaborated into an amazing variety of foods, and the processing is particularly elaborate. Of concern here is the most common manioc harvesting and preparation routine in which raw manioc tubers are transformed into a boiled juice (*manicuera*), starch drinks (*mingao*), and bread (*casabe*). Manioc bread and one of the drinks form part of almost every ordinary meal in the village.

The daily manioc routine can be divided into three phases: acquisition, processing and cooking, which are shown schematically in Figure 3. This diagram can be usefully compared with the more detailed presentation of Hugh-Jones (1979:174–180).

Acquisition

The acquisition phase takes place away from the village. It involves travel to the garden, harvesting, related work and the transport of produce from the garden (Figs. 4 and 5). Peeling and washing are included in the acquisition phase, since the tubers are usually peeled in the garden and washed in a stream on the return trip to the village.

Processing

Because manioc tubers deteriorate very rapidly, they are usually processed the same day they are harvested. Processing improves the storage characteristics of the starch products, as well as decreasing the level of toxicity.

In processing the tubers are separated into liquids, starch and fiber. This is accomplished by grating and sieving. The tubers are first grated on a wooden grating board set with stone chips (Fig. 6). The grated mash is then worked in a large basketry sieve to express the liquids. Water is added to wash out the starch. The starch and liquids are collected under the sieve in a large pot; the starch is allowed to settle and the liquids decanted off the top (Fig. 7).

Once separated the starch and the fiber are relatively stable and can be stored in leaf-lined baskets or floor pits for several weeks or more. Under these storage conditions they ferment slightly, and it is in this

Figure 4. Woman harvesting manioc tubers. Women typically remove the top portion of the woody stalk and then pull up the tubers. (Photo by Paul N. Patmore.)

Figure 5. Woman carrying a morning's harvest of unpeeled manioc tubers. This is a relatively large harvest intended for the production of manioc beer. It is approximately twice the amount harvested to meet daily households needs. (Photo by Paul N. Patmore.)

Figure 6. Woman grating manioc tubers on a traditional wooden grating board set with stone chips. A basket of unpeeled tubers is on her left. Women grate with long, rhythmic strokes, using both hands simultaneously. A typical daily harvest of 15 kg of tubers can be grated in about 1 hour. (Photo by Paul N. Patmore.)

Figure 7. Woman sieving grated manioc mash. The liquids and starch expressed from the mash are being collected in the large pan below the sieve. Later, when the starch has settled, the liquid will be decanted and boiled as a drink. (Photo by Paul N. Patmore.)

fermented form, rather than fresh, that they are used. Raw manioc juice, on the other hand, is very unstable and cannot be stored.

Cooking

Raw manioc juice is considered very toxic and is always boiled immediately. The boiling volatilizes the toxic hydrocyanic acid (HCN) liberated in processing, and produces a slightly sweet beverage. The serving of boiled juice in the late afternoon or early evening marks the end of the daily manioc routine. The boiled juice cannot be stored for any length of time, and in rare instances, when it is not consumed within 16 to 18 hours, it spoils and is discarded.

Starch drinks are prepared by dissolving some slightly fermented starch in boiling water and allowing it to cook until thickened. These drinks are sometimes flavored with palm fruits, bananas, pineapple or lemon and are prepared for meals when manioc juice is not available.

Manioc bread can be prepared in a number of ways, but in Yapú one form is particularly characteristic: a thick, soft bread made by

Figure 8. Woman toasting *fariña* on a clay griddle. Fariña is a dry meal made from yellow manioc tubers which have been soaked in stream water for 2 or 3 days, peeled, grated and then allowed to ferment. (Photo by Paul N. Patmore.)

recombining the fermented fiber and starch. The fiber is first squeezed dry in a *tipiti* (a diagonal weave basketry sleeve) and lightly toasted. Moist starch is then added to the pre-toasted fiber and the mixture baked on a clay griddle in the form of a large round bread.

The Tatuyo prefer fresh bread daily. Although the bread could possibly be dried and stored for long periods, this is not done. Rather, when a carbohydrate source with good storage characteristics is needed, *fariña* is used. Fariña, a dry manioc meal, is a light weight, concentrated source of calories, and is the form in which manioc is carried when traveling. (Fig. 8)

The phases of acquisition, processing and preparation correspond

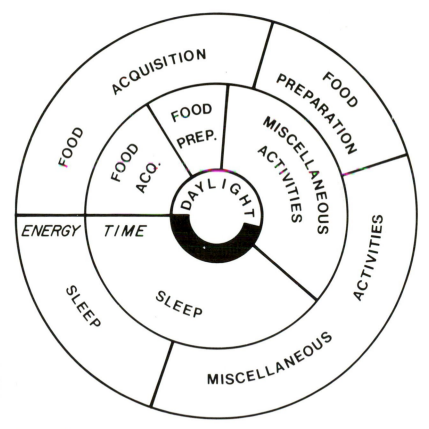

Figure 9. Mean daily time and energy budgets of Yapú women expressed as a percentage of total time per day (24 hours), and total mean daily energy expenditure.

to the way in which women's work is organized. Women ordinarily leave for the garden early in the morning. They return to the village around mid-day, rest and eat. The afternoon is largely taken up with manioc processing. Firewood is collected in the late afternoon and fresh manioc juice boiled. Manioc bread is usually baked in the late afternoon while the juice is being boiled.

TIME-ENERGY BUDGETS OF WOMEN

One gets the impression that the daily manioc routine takes up most of women's time. This impression is substantiated by an examination of time-energy budgets. A time-energy budget is the way in which time and energy are allocated to various activities; that is, the way in which time is spent and energy expended.

The mean daily time-energy budget of four Yapú women is shown schematically in Figure 9 (See also Table I). The time and energy expenditure of each woman was monitored for a 3-day period, hence

TABLE I

Percent mean time and energy budgets of four adult Yapú women during three days as a percentage of total time per day, total daylight time and total mean daily energy expenditure

Activity	Total Time	Energy Expenditure	Daylight Time
Food Acquisition			
Travel	6.0	11.7	11.0
Gardening	9.6	16.9	17.8
Collecting	0.3	0.6	0.6
Food Preparation			
Processing	3.2	6.1	5.9
Cooking	5.5	6.9	10.2
Fuel and water acq.	1.6	2.8	2.9
Subtotal	26.2	45.1	59.4
Miscellaneous Activities	34.6	35.8	
Sleep	39.2	19.2	
Total	100.0	100.0	

NOTE: Total time per day is 24 hours; total daylight time is considered 13 hours; total mean daily energy expenditure for these women was 2098 kcal.

the values are the means of 12 woman-days. The women gardened on 10 out of the 12 days. Of these, 8 were routine manioc harvesting and processing days. This pattern of gardening and non-gardening days is representative.

The time budget is expressed as a percentage of the total time per day (24 hours). Similarly, the energy budget is expressed as a percentage of the total mean daily energy expenditure, which for these women was 2908 kcal. Both are divided into four major categories: food acquisition, food preparation, miscellaneous activities in the village and, finally, sleep. Of particular interest are the categories of food acquisition and food preparation, which together account for 26% of the total mean daily time budget and 45% of the total mean energy budget.

The significance of the time budget is best understood in relation to available daylight. Darkness is a constraint on activity, and women complete all food acquisition[5] activities and most food preparation by dusk. Since daylight in the tropics is about 13 hours, food acquisition and preparation account for 59.4% of the total daylight time available to women.

Food acquisition includes travel to and from resource areas, gardening and collecting. Travel to and from gardens accounted for almost half of all the time and energy labeled food acquisition, or about 6% of the total time budget, and 11.7% of the total energy budget. It averaged almost two hours (115 minutes) per garden trip for these women. This travel time includes some resting along the trail and bathing on the return trip.

Gardening includes all activities that take place in the garden related to the harvesting and maintenance of cultigens. Although manioc harvesting and the associated weeding and replanting are the dominant activities, it also includes the harvesting of other products, fire building and childcare. Gardening accounts for about 9.6% of the time, and 16.9% of the mean daily energy budget. If only the 10 days on which the women actually gardened are considered, the values are slightly higher, 10.7% (2 hours and 34 minutes) and 18.7% (393 kcal) of the mean time and energy budgets respectively.

At this time of the year, the gathering of wild vegetable products and small invertebrates is relatively unimportant. In the 12 woman-days considered here, the women only collected on two occasions. The first was a short trip from the village to gather ants, and the second a short detour from the garden trail to search for a wild fruit, *poá* (*Parkia* sp.).

The category of food preparation accounted for about 10.3% of the total time and 15.8% of the total energy budget. It includes both the extensive processing of manioc and the cooking of manioc products and other foods. Within this category, manioc processing accounts for about a third of the time and energy expended, and cooking for about half. Fuel and water acquisition used a relatively small proportion of the total time and energy budget, 1.6% and 2.8% respectively. Both of these resources were very accessible.

The time and energy devoted to cooking is underestimated in this study because of the nature of the activity and the coding system used. A general code of miscellaneous household activities was employed when a variety of activities were undertaken within a short period of time. Thus, some cooking, when interspersed with other activities, was coded under the miscellaneous category.

It should be noted that the days of observation included only routine manioc harvesting and processing. The making of beer from manioc, which is a fairly frequent activity, involves a process even more lengthy than the one sketched here. The preparation of *fariña* (manioc meal) is also very time consuming.

The time and energy devoted by Yapú women to gardening and food preparation appear greater than in some other tropical forest groups. Table II compares the time and energy inputs of Yapú women in gardening and food preparation with those of the Machiguenga of Southeastern Perú, and New Guinea women from the villages of Kaul,

TABLE II

Percent mean daily time (t.) and energy expenditure (e.e.) of Yapú women in gardening and food preparation compared with the Machiguenga of Southeastern Peru, and with women from three villages in New Guinea (Kaul, Lufa and Pari)

Site/Ethnic Group	Gardening		Food Preparation	
	t.	e.e.	t.	e.e.
Yapú	9.6	16.9	10.3	15.8
Machiguenga[a]	3.6	7.0	9.7	13.9
Kaul[b]	2.8	5.3	5.5	6.4
Lufa[b]	6.4	12.2	3.0	2.5
Pari[c]	7.7	14.7	—	—

[a] Data from Montgomery and Johnson (1977:104): mean of wet season and dry season values.
[b] Data from Norgan, Ferro-Luzzi and Durnin (1974:328).
[c] Data from Hipsley and Kirk (1965:100). Some gardening done on all days.

Lufa and Pari. All of these groups depend on rootcrop horticulture for a significant portion of their energy intake. It is clear that the Yapú women spend more of their time and energy budgets in both gardening and food preparation than do the New Guinea women. The difference is especially striking in the energy expended in food preparation. Yapú women spend two to five times more of their energy budget in this activity than do those from Kaul and Lufa, respectively.

These differences in time and energy expenditure in food preparation are a reflection of differences in the processing and cooking of basic staples. Over 80% of the calorie intake in Kaul comes from vegetables (taro, 42%), fruit and store foods (Norgan, Ferro-Luzzi and Durnin 1974:323), which probably do not require elaborate processing. The pattern in Lufa is similar, with vegetables (sweet potatoes 63.8%) and store foods accounting for over 80% of the calorie intake (Norgan, Ferro-Luzzi and Durnin 1974:323). In Yapú, on the other hand, store foods are not a part of the diet, and the bulk of the calories are derived from manioc which requires elaborate processing. The Machiguenga also rely on manioc as a dietary staple but prepare much of it as beer, using a simpler method than is used in Yapú. They devote more of their time and energy budgets to food preparation than do New Guinea women, but somewhat less than do Yapú women.

CONCLUSIONS

Domestic groups in Yapú strive to be energy and nutrient self-sufficient. The reliance on manioc as a principal source of energy, and the manner in which production is socially organized, require consistent and relatively high inputs of the time and energy of women. The day-to-day work of women is crucial to the maintenance of domestic group self-sufficiency because the food energy they supply constitutes a large portion of the total and because they maintain a consistent rate of energy flow in the face of extreme fluctuations in other energy sources.

There are two further implications of this reliance on manioc. First, anything that interferes with the inputs of time and energy necessary to sustain manioc production threatens the self-sufficiency, and hence, integrity of the domestic group. Such interference is commonly the result of illness or disability. Second, the maintenance of stable energy flow by women allows men considerable flexibility in their time-energy budgets, a flexibility which is critical to the exploitation of available animal protein resources.

NOTES

1. Financial assistance for the field work in the Vaupés was provided by a Dissertation Fellowship from the Social Science Research Council, a grant from the National Science Foundation (No. BSN 75-20169) and a Fellowship from the Organization of American States. Grateful acknowledgement is made to Paul N. Patmore for the preparation of the graphic materials and to Catherine LeGrand for her helpful suggestions.
2. The distinction made between bitter and sweet manioc adheres to native categories. In this paper the word manioc is used to refer only to bitter varieties of the crop, which are considered toxic and are the dietary staple. Sweet manioc is of minor importance to the Tatuyo and is used like other rootcrops such as yams.
3. All of the food energy in this category was derived from *bati* (*Erisma Japura*), an oil seed harvested the previous May and June. It is processed into a butter and stored in leaf-lined pits for as long as a year. (See Dufour and Zarucchi 1979.)
4. Dry coca leaves contain some 305 Cal and 18.9 g protein per 100 g, as well as significant amounts of other nutrients (Duke, Aulik and Plowman 1975:199). Thus, an intake of 50 g/day, which was common for some men, provided significant amounts of calories and protein.
5. There is only one exception to this: the collection of ants (*Atta cephalotes*) during their pre-dawn nuptial flights.

REFERENCES

Coursey, David G. 1973. Cassava as food: Toxicity and technology. pp. 27–36 in *Chronic Cassava Toxicity: Proceedings of an Interdisciplinary Workshop*, ed. by Barry Nestel & Reginald MacIntyre. Ottawa: International Development Research Center

Dufour, Darna L. 1981. *Household Variation in Energy Flow in a Population of Tropical Forest Horticulturalists*. Ph.D. dissertation, State University of New York at Binghamton

_____ 1983. Nutrition in the northwest Amazon: Household dietary intake and time-energy expenditure. pp. 329–355 in *Adaptive Responses of Native Amazonians*, ed. by Raymond B. Hames & William T. Vickers. New York: Academic Press

Dufour, Darna L. and James L. Zarucchi. 1979. *Monopteryx angustifolia* and *Erisma Japura*: Their use by indigenous peoples in the Northwest Amazon. *Botanical Museum Leaflets* 27(3–4): 69–91

Duke, James A., David Aulik and Timothy Plowman. 1975. Nutritional value of coca. *Botanical Museum Leaflets* 24(6): 113–119

Hipsley, Eben H. and Nancy E. Kirk. 1965. *Studies of Dietary Intake and Expenditure of Energy by New Guineans*. South Pacific Commission Technical Paper No. 147. Noumea.

Hugh-Jones, Christine. 1979. *From the Milk River: Spatial and Temporal Processes in the Northwest Amazon*. London: Cambridge University Press

Jones, William O. 1959. *Manioc in Africa*. Stanford: Stanford University Press

Montgomery, Edward and Allen Johnson. 1977. Machiguenga energy expenditure, *Ecology of Food and Nutrition* 6:97–105

Norgan, N. G., A. Ferro-Luzzi and J. V. G. A. Durnin. 1974. The energy and nutrient intake and the energy expenditure of 204 New Guinean adults. The Royal Society. *Philosophical Transactions B*. Biological Sciences 268(893):309–348. London.

ADDITIONAL READING

NORTHWEST AMAZON: GENERAL

Arhem, Kaj. 1976. Fishing and hunting among the Makuna. Göteborg Etnografiska Museum. *Arstryck* (Annals) 1976:27–44. Gothenburg, Sweden
Bidou, P. 1976. *Les Fils de l'Anaconda Céleste (Les Tatuyo): Etude de la Structure Socio-Politique*. unpublished Ph.D. dissertation, Troisieme Cycle, University of Paris
Goldman, Irving. 1963. *The Cubeo*. University of Illinois Studies in Anthropology #2. Urbana, Il.: University of Illinois Press
Hugh-Jones, Stephen. 1979. *The Palm and Pleiades: Initiation and Cosmology in Northwest Amazonia*. London: Cambridge University Press
Jackson, Jean E. 1976. Vaupés marriage: A network system in an undifferentiated region of the Northwest Amazon. pp. 65–93 in *Regional Analysis: Social Systems*, ed. by Carol Smith. New York: Academic Press
_____ 1983. *The Fish People: Linguistic Exogamy and Tukanoan Identity in Northwest Amazonia*. New York: Cambridge University Press
Reichel-Dolmatoff, Gerardo. 1971. *Amazonian Cosmos: The Sexual and Religious Symbolism of the Tukano Indians*. Chicago: University of Chicago Press

MANIOC AND MANIOC PROCESSING

Albuquerque, Milton de 1969. *A Mandioca na Amazônia*. Belém: Superinten-dência do Desenvolvimento da Amazônia
Cock, James H. 1982. Cassava: A basic energy source in the tropics. *Science* 218:755–762
Dole, Gertrude. 1978. The use of manioc among the Kuikuru: Some interpretations. pp. 217–247 in *The Nature and Status of Ethnobotany*, ed. by Richard I. Ford. Museum of Anthropology, Anthropological Papers No. 67. Ann Arbor, Mich.: University of Michigan
_____ 1969. Techniques of preparing manioc flour as a key to culture history in tropical America. pp. 241–248 in *Men and Cultures: Selected Papers of the Fifth International Congress of Anthropological and Ethnological Sciences*, ed. by Anthony F. C. Wallace. Philadelphia: University of Pennsylvania Press
Lancaster, P. A., J. S. Ingram, M. Y. Lim and David G. Coursey. 1982. Traditional cassava-based foods: Survey of processing techniques. *Economic Botany* 36:12–45
Moran, Emilio. 1973. Energy flow analysis and the study of *Manihot esculenta* Crantz. *Acta Amazonica* 3(3):29–39
_____ 1975. Food, development and man in the tropics, in *Anthropology of Food and Food Habits*, ed. by Margaret L. Arnott. The Hague: Mouton

TIME AND ENERGY BUDGETS IN TRIBAL SOCIETIES

Johnson, Allen W. 1975. Time allocation in a Machiguenga community. Ethnology 14:301–310
Johnson, Orna R. and Allen W. Johnson. 1976. Male-female relations and the organi-

zation of work in a Machiguenga community. *American Ethnologist* 2:634–648
Richards, Audrey I. 1939. *Land. Labour and Diet in Northern Rhodesia: An Economic Study of the Bemba Tribe*. London: Oxford University Press

WOMEN'S WORK IN TRIBAL SOCIETIES

Dahlberg, Frances. 1981. *Woman the Gatherer*. New Haven, Conn.: Yale
Murphy, Yolanda and Robert F. Murphy. 1974. *Women of the Forest*. New York: Columbia University Press
Siskind, Janet. 1973. *To Hunt in the Morning*. New York: Oxford University Press

CHAPTER 2

RICE IN A LUA' SUBSISTENCE ECONOMY, NORTHWESTERN THAILAND[1]

PETER KUNSTADTER
East-West Center
Honolulu, Hawaii 96848

ABSTRACT

Rice is the major agricultural product and source of food energy for Lua' subsistence farmers in the northwestern Thailand hills. Since the mid-1960s, responding to land shortages and population growth, the villagers have shifted emphasis from swidden to irrigated rice. Although only traditional methods were used, total rice production and production per worker increased by 44% between 1967 and 1979. Because of population increase, production *per capita* increased only 3.5% during this period.

Energy produced *per capita* in the form of rice rose from 1.25×10^6 Cal in 1967 to 1.29×10^6 in 1979. In 1968 villagers ate an average of 710 g (uncooked rice) per day (2618 Cal). A survey in 1980 showed that villagers ate a mixed diet, with vegetables as the major supplement to their rice staple in 62% of their meals, animal protein in 27%, or only peppers and seasoning in 12%. Young children were often given extra meals supplemented with vegetables or animal protein.

The economic situation of these villagers appears to have improved with more wage work opportunities and increased rice production despite rapid growth of village population in the 1970s.

INTRODUCTION

This paper describes changes between the late 1960s and 1980 in rice production and consumption in the Lua' village of Pa Pae, in the hills of Mae Sariang District, Mae Hongson Province, northwestern Thailand.[2,3] The economy and social organization of the Pa Pae

21

villagers has traditionally been based on swidden cultivation of non-glutinous rice (*Oryza sativa*) using a relatively stable, conservative forest fallow system. Many other crops are grown in the swiddens and in creek-side gardens which are cultivated during the cold season. Villagers supplement their diet with fruit from a number of varieties of trees planted around the village and with ferns, mushrooms and many other vegetable products which grow wild in the fallow swiddens and uncut forest (Kunstadter, Sabhasri and Smitinand 1978; Kunstadter 1978a, 1978b). Villagers began making irrigated fields in the early 1930s, but only in the past decade has irrigated rice gained preeminence in the local economy.

Traditionally the villagers owed feudal allegiance to Northern Thai princes and paid them a small annual tribute in exchange for access to land and the right to govern their own village affairs (Nimmanahaeminda 1965). The feudal relationship was broken when the Royal Thai Government took control of the north around the turn of the century and imposed heavy head taxes payable in cash. This pushed the villagers into the wage labor market and increased pressure for sales of farm and forest products. Internal village affairs were largely left in the hands of village leaders for the first six decades of Thai administration. Beginning in the early 1960s, central government programs were introduced to hill villages, including Pa Pae. These started with schools and a little medical care. In the 1970's a series of government or quasi-government economic development projects were started at Pa Pae, mostly by introducing cash crops. Some of these projects also provided wage work opportunities for villagers who previously had to look for work in the lowland valleys or in mines a day or more distant from the village.

Other changes, especially rapid population growth in Pa Pae and neighboring villages, loss of traditional village lands to neighboring villages, conversion of over half the villagers from animism and Buddhism to Catholicism or fundamentalist Protestantism, and the construction of a dry season road to Pa Pae, have weakened the traditional social organization, leading to less careful and conservative community management of swiddening since the late 1960s.

Village life still revolves around rice production, with the cycle of activities primarily regulated by the monsoonal climate and the requirements of swidden agriculture. Almost all adults are still primarily farmers, and almost all food is still locally produced. Some members of households also find it necessary to engage in wage work, since farming gives them little cash income, and by now they have acquired needs for a wide variety of modern market goods, some of which have replaced traditional village crafts.

SWIDDEN CULTIVATION

Swiddening is done according to a regular field rotation system, with one year of cultivation followed by eight years of fallow. The length of fallow was reduced from nine to eight years in 1978 when one of the annual swidden blocks was taken over by a neighboring village. It is still too soon to tell the effect of reducing the fallow period on what had been a reasonably stable system that allowed forest to regenerate, and maintained soil fertility on fields which had been regularly swiddened for many generations (Kunstadter 1978; Zinke, Sabhasri and Kunstadter 1978; Sabhasri 1978).

There is little obligatory division of swidden labor by sex. Both men and women perform most tasks; both sexes cut down small trees when clearing swiddens. Men rather than women climb large trees to lop off branches which might shade swidden crops; men wield the long iron tipped digging sticks to make holes for swidden rice seed. Both men and women do the actual planting and both sexes weed, but women do more weeding than men. Men are more likely to leave the village to seek wage work during the weeding season. Both sexes harvest rice, but more often it is men who thresh by beating the sheaves on bamboo mats, while women carry the sheaves to the threshing floor. Both sexes carry the rice back to the village after harvest. Youngsters assist in this work as they grow large enough to help.

Villagers cut coherent blocks of swidden fields in January or February, burn the slash in March, plant in April, weed from May to September and harvest in October and November. They use only hand tools, with minimal disturbance of the soil surface during planting and weeding. Forest regrowth is rapid from the coppiced stumps of fire resistant tree species which are left when the field is cut, burned and cultivated. Grassland often replaces forest in swidden areas in Thailand (and elsewhere in the tropics). This has not occurred under the Lua' field rotation swidden system, probably because of rapid regrowth of forest species from stumps that the Lua' farmers leave in the swiddens, and through their control of fire during the fallow period. In fact, *Imperata* grass, which is used for roofing straw, is becoming scarce. This is probably a result of increased grazing pressure since many of the irrigated fields have been fenced off for a second crop during the dry season, and since increasing numbers of water buffalo and cattle are being raised in the village.

Traditionally swidden land was communally owned by the descendants of village founders and could not be sold or otherwise disposed of except by village leaders on behalf of the whole village. Use rights to particular plots were generally maintained by households from one cultivation cycle to the next, but swidden sites could be reassigned by the village religious leaders to whom the villagers owed allegiance. Plots were reassigned to adjust for changes in household size or the creation of new households, or in the case of disputes between village households over a single plot of land.

The authority of traditional animist religious leaders to control land use was based on their ability to appease the spirits which were believed to own or live in the swidden areas. They did this by calling the spirits to partake in animal sacrifices to which each household was obliged to contribute. Recently the religious leaders have been losing control over land as individual households have rented or sold their plots to outsiders, and as swidden land has been converted to privately owned irrigated fields. Traditional swiddening required the coordinated efforts of all village households to burn the swidden block and prevent the accidental spread of fire. The date of the fire was set by the village religious leaders, and the burn was accompanied by village-wide ceremonies. Other village-wide ceremonies were held annually in connection with swidden weeding (to appease the spirit owners of the fields) and after the harvest had been brought home (to call the souls of people and of the rice back from the swiddens to the village). As the civil law of the Thai administration has supplanted customary law, and as more villagers have been converted to Christianity, village leaders are less able to enforce their rule through animist religious sanctions. The observance of communal ceremonies has declined as Christian households refuse to contribute to the purchase of sacrificial animals.

In the past the major constraint on swidden rice production was the requirement for labor during the long tedious weeding season. Good rice production depends on the elimination of competing weeds. A second constraint which has become much more important in recent years is the scarcity of good quality land, for which the competition from neighboring villages is increasingly heavy. Another limitation on production is the control of fires in fallow swiddens, which slow the restoration of soil fertility. Fires traditionally were strictly controlled by village leaders who determined the time and place of swidden fires and required that each household contribute one person's labor whenever a fire started or escaped in fallow areas. As the population of nearby villages has expanded and the emphasis on swiddening has declined (see below), the villagers and their leaders find it ever more difficult to prevent and control accidental fires, and even to coordinate the timing of swidden fires.[4]

IRRIGATED FARMING

Irrigated fields which have been terraced in the well-watered gradually sloped valley lands are privately owned and can be bought, sold and mortgaged without permission from village leaders. Coordination of work in the wet rice fields is arranged between the several owners whose fields are fed by a common ditch. No particularly critical moment occurs in the wet rice cycle. Operations are timed by individuals, depending on whether water is available on their fields and labor and draft animals are available for field preparation. Men and youths do most of the work of preparing ditches and fields, but if necessary, girls rather than boys may lead the water buffalo when plowing and harrowing. Both sexes participate in transplanting, weeding and harvesting.

Rice nurseries are normally sown in May or June and may either be irrigated or swiddened. The fields are plowed and harrowed after heavy rains have softened them, usually in June. Ditches, dams and dikes are repaired any time before the heavy rains cause streams to rise, usually early in June. Fields are transplanted when the seedlings are big enough and water is available to flood the fields, usually starting late in June and lasting through much of July. Unlike the swiddens, which must be planted after the fields are burned and before the rains start, there is no particular pressure to finish transplanting by a certain date. Two weedings are carried out, but they are not very onerous, as flooding and drying controls most of the weeds. Harvest begins in October when the rice ripens, often slightly later than the swiddens. New fields may be dug and old fields expanded at any time from after the harvest to beginning of the heavy rains. This means that except for transplanting and harvest, which may be delayed for a few days without much damage, there is no requirement for concentrated or even sustained labor. The major constraints on irrigated rice production are access to irrigated land and draft animals (i.e. capital investments or rental and water).

Availability of large amounts of non-family labor for brief periods of concentrated work is less of a constraint than in swiddens. Unlike swiddens, which are fallowed after one rainy season of cultivation, the irrigated fields are used every year, and the potential for dry season cropping is now being explored with soy beans intended as a cash crop. The sale of cash crops outside the village is difficult because the cost of transportation is high and it is hard to get the soy beans to market.

RICE PRODUCTION

Observations in the late 1960s and 1979-80 show the extent to which rice production has intensified through increased use of irrigation. Amounts of rice produced in swidden and irrigated fields are shown in Table I. While volume of rice planted (an approximate measure of area planted) and swidden yields remained practically constant over the past decade, the amount of irrigated rice planted has increased 58% and the amount harvested has increased about 90%. Swidden yields per unit of seed have remained about the same (about 11:1) while irrigated rice yields have increased about 20% to 32:1 (Table II).

Increase in total volume of production has slightly exceeded the 2.74% average annual growth in population. The number of people increased from 221 in 1967 to 316 in 1980, while the number of workers

TABLE I

Rice production at Pa Pae village (liters, unhusked)

	1967		1968		1979		1980	
	Seed	Yield	Seed	Yield	Seed	Yield	Seed	Yield
Swidden								
Early	1205	11660	1470	13220	786	8400	500	3860
Late	7800	89840	9890	103560	9350	100920	8194	94648
Glutinous	363	3680	495	5465	687	9920	707	12762
Total	9368	105180	11855	122245	10823	119240	9401	111270
Irrigated								
Non-glutinous	2680	74040	2605	68960	4385	138000	4026	129772
Glutinous	51	620	122	580	61	1355	142	4025
Total	2731	74660	2727	69540	4446	139355	4168	133797
Swidden + Irrigated		179840		191785		258595		245067

(healthy persons working or able to work regularly in the fields) increased only slightly, so that the ratio of dependents to workers almost doubled, from 0.75 to 1.23. Despite the increase in the proportion of nonworkers (primarily young children), the volume of rice produced per person in the population has remained about the same. This has been made possible by the increase in production per worker at an annual rate of 2.00% (average for 1967–1968 vs. 1979–1980) (Table III).

Rice production is dependent primarily on household labor and on reciprocal labor exchange for work contributed to other households. There is almost no wage labor in Pa Pae agriculture. Swiddens are made by village households on village-controlled land. Irrigated fields are held in fee and are almost always worked by their owners. Only a few irrigated fields (owned by widows and by one man who has more fields than he and his son can work) are rented out. Labor supply for household farming is affected primarily by manipulation of household size through marriage and reproduction rather than by adoption or hiring of non-family members.[5] As already noted, reciprocal labor exchange is important during times of maximum labor demand (transplanting and harvesting).

The ideal household life cycle is for the eldest son to marry and bring his wife to stay with his parents. This couple will build a new house

TABLE II

Yield per unit of seed (by volume, unhusked) at Pa Pae village

	1967	1968	1979	1980
Swidden	11.23	10.31	11.02	11.84
Irrigated	27.34	25.50	31.34	32.10

TABLE III

Population and rice production (liters, unhusked) at Pa Pae village

	1967	1968	1979	1980
Persons	221	233	307	316
Workers	126	124	126	142
Households	50	51		
Dependents/Workers	0.75	0.88	1.43	1.23
Rice Produced/Person	813.8	823.1	842.3	775.5
Rice Produced/Worker	1427.3	1546.7	2052.3	1725.8

when the man's younger brother marries. Matrilocal marriages occur when the groom wants access to land controlled by the bride's household, and when the bride's parents have no surviving son (Kunstadter 1966b). Parents are often anxious for their sons to marry and to bring another able-bodied worker into the household. This is only a temporary advantage, because mothers of young children are often unable to work in the fields unless their children can be cared for by the grandparents. Young married couples with dependent children are often relatively poor and are in special difficulties if they have no parents with whom to leave their children while they work the fields. The most fortunate families are those with several unmarried adult children still living at home. Not only do these children contribute to the household work force, but they also attract other bachelors and maidens to their fields at times of maximum labor input (planting, harvesting), which are important occasions for courtship. Attracting a large labor force for concentrated labor at these times is one of the major managerial tasks for household production units which must rely on bonds of kiship and reciprocal labor exchange to get the work done. Thus working-age children (age about 12 and over) are a major asset above and beyond the labor they contribute to the households.

Despite the need for household labor it is now clear that parents do not want an unlimited number of children, especially young, dependent ones. Half the married women between age 30 and 44 were using birth control as of 1980, and many express the worry that they will not be able to support their children. Though they are increasingly aware of wage

TABLE IV

Correlations (r^2): Total yield (volume) of Pa Pae rice production per household with number of persons and number of workers per household

	1967	1968	1979	1980
Total household yield and persons in household[a]	0.166	0.184	0.127	0.198
Total household yield and workers in household[b]	0.267	0.361	0.279	0.260
Total household yield and workers/persons in hh[c]	0.000	0.003	0.045	0.000

[a] Probability of all correlations in this row less than 0.001
[b] Probability of all correlations in this row less than 0.0001
[c] Probability of all correlations in this row more than 0.05

work or professional work opportunities outside the village, the majority of village parents still want and expect their children to live in the village and help with farming when they marry.

Correlation of total household rice production with variations in household composition and size (total household size, number of workers per household, ratio of workers to total number in household) suggests that the most important factor in determining household productivity is number of workers per household (r^2 ranging from 0.260 0.361). This correlation is considerably stronger than is the correlation of total household rice production with number of people in the household (r^2 ranging from 0.127 to 0.198) (Table IV). This suggests that farm workers are producing as much as they are able, not just matching production to the number of consumers in the household, contrary to what Chayanov (1960) suggests in the "law of least effort".

ANALYSIS OF CALORIE CONTENT OF RICE PRODUCED BY PA PAE FARMERS

Composition of samples of swidden and irrigated rice, and calculation (by Atwater's method) of calories in husked, uncooked Pa Pae rice is shown in Table V. The calculations show no significant difference in the varieties analysed as regards calorie content, which is about 355 Cal per 100 g of husked rice. There is considerable variation in nutrient composition in different local rice types. Total protein, for example, ranged from 6.93 to 8.65% of the weight of oven dried husked rice in six different types of Pa Pae rice[6], (i.e., one variety had about 25% more protein by weight than did the other). This is important because rice is a major source of protein as well as the primary source of calories in the Pa Pae diet. Differences in protein content could be particulary significant for growing children whose needs for protein are higher and whose rice consumption is smaller than among adults.

In order to calculate the number of food calories in rice produced at Pa Pae, we have converted the volume of unhusked rice produced to the weight of husked rice. Conversion factors and calculation of total food calories in the rice produced are shown in tables VI and VII. In husked form swidden rice is about 2% heavier per unit volume than is irrigated rice (826.5 vs. 809 g/l). About half the field volume of rice is lost during the husking process, which traditionally is done with a foot-powered pestle and large wooden mortar. Two diesel-powered mills were brought to Pa Pae by village entrepreneurs in 1979 (aided

TABLE V

Proximate composition of husked swidden and irrigated Pa Pae rice[a] and food Calories per 100 g calculated by Atwater's Method[b]

Rice Type	Moisture	Protein	Fat	Carbohydrate	Fiber	Ash	Total	Average Calories
Swidden								
Non-glutin. pounded								
Phuan Lawng Chuang	12.49	8.65	1.64	76.19	0.12	0.91	100.00	
Calories	—	29.50	13.73	313.91	—	—	357.14	
Non-glutin. milled								354.245
Phuan Lawng Chuang	13.03	6.93	1.06	78.45	0.10	0.43	100.00	
Calories	—	23.64	8.87	323.22	—	—	355.73	
Glutinous pounded								
Phit Phae	14.40	7.08	1.08	76.74	0.15	0.55	100.00	
Calories	—	24.15	9.04	316.17	—	—	349.36	
Glutinous pounded								
Phit Sakyak	13.08	7.04	1.42	77.39	0.39	0.68	100.00	
Calories	—	24.01	11.89	318.85	—	—	354.75	
Irrigated								
Non-glutin. pounded								
Cheuk Lawng Lathin	12.32	7.54	1.19	78.15	0.34	0.46	100.00	
Calories	—	25.72	9.96	321.98	—	—	357.66	356.37
Non-glutin. milled								
Cheuk Lawng Lathin	12.98	8.42	1.27	76.73	0.11	0.49	100.00	
Calories	—	28.72	10.63	316.13	—	—	355.48	

[a] Analysed by Institute of Food Research and Product Development, Kasetsart University, Bangkhen, Bangkok 9, Thailand.
[b] Calculated from values for "Rice, brown," in Table 6, U.S. Dept. of Agriculture Handbook 8 (1963:160).

by lowland investors). These deliver approximately the same rate of milled rice per volume of unhusked rice as does foot pounding, but at a charge to the customers of one liter of milled rice per 20 l brought for milling, plus the chaff. the machine milled rice appears to be slightly more highly polished than the foot-pounded rice, but there

TABLE VI

Volume and weight changes during the husking of Pa Pae rice

	Weight of Unhusked Rice (g/l)	Loss in Volume during Milling	Weight of Foot-Pounded Rice (g/l)	Conversion Factor for Unhusked Volume to Husked Weight
Swidden	620	.5	826.5	.41325
Irrigated	540	.5	809.0	.40450

TABLE VII

Calculation of food Calories in rice produced at Pa Pae

	1967	1968	1979	1980
Swidden Rice Unhusked volume (1)	105,180	122,245	119,240	111,270
Husked weight (g) @ 413.25 g per unhusked liter	43,465,635	50,517,746	49,275,930	45,982,328
Calories @ 354.25 per 100 g of uncooked husked rice	153,977,000	178,959,000	174,559,000	162,892,000
Irrigated Rice Unhusked volume (1)	74,660	69,540	139,355	133,797
Husked weight (g) @ 404.5 g per unhusked liter	30,199,970	28,128,930	56,369,097	54,120,886
Calories @ 356.57 per 100 g of husked rice	107,684,000	100,299,000	200,995,000	192,978,000
Swidden & Irrigated Total food Calories	261,661,000	279,258,000	375,554,000	355,870,000

TABLE VIII

Food Calories in the form of rice produced at Pa Pae per person and per worker, per year and per day

	1967	1968	1979	1980
Total calories produced in swidden and irrigated rice				
Per person per year	1,183,986	1,198,532	1,223,303	1,126,171
Per worker per year	2,076,675	2,252,080	2,980,588	2,506,127
Per person per day	3,244	3,284	3,356	3,085

seems to be little effect of milling method on amino acid content (see below).

Total food calories in the rice produced anually at Pa Pae have increased from 270.5 million in 1967–1968 to 366 million in 1979–1980. The increase is primarily from the additional irrigated rice planted and produced. Two-year average calorie production per worker increased 35% from 1967–1968 to 1979–1980, but, as suggested by figures on volume of rice produced, the number of rice calories produced *per capita* has not increased (Table VIII). The lack of *per capita* increase is due primarily to the growth of population size (and decreased proportion of workers in the population associated with the change in age structure which has accompanied rapid population growth) and, secondarily, the increased proportion of irrigated rice, which has a slightly lower weight per unit volume than does swidden rice.

AMINO ACID CONTENT OF SEVERAL RICE VARIETIES

The amino acid contents of several rice varieties from Pa Pae are given in Table IX. Varieties selected for comparison were as follows: one of the most common swidden non-glutinous types, milled by foot-powered pestle and mortar, Lua' name *hngaw phuan lawng chuang* (*Oryza sativa*) the same variety, from the same farmer, milled by machine; a swidden glutinous rice variety selected for its ability to grow on poor soil, milled by foot pounding, *hngaw phit sakyak* (*O. sativa* var. *glutinosa*), a more common variety of swidden glutinous rice, milled by foot-pounding, *hngaw phit phae*; and a common

TABLE IX
Amino acids in five rice samples from Pa Pae, 1980[a]

| | Swidden, Non-glutinous, Foot Pounded | | Swidden, Non-glutinous, Machine Milled | | Swidden, Glutinous, Poor Soil, Foot Pounded | | Swidden, Glutinous, Ordinary Soil, Foot Pounded | | Irrigated, Non-glutinous, Foot Pounded | |
| | g a.a.[b] | mg a.a.[c] | g a.a.[b] | mg a.a.[c] | g a.a.[b] | mg a.a.[c] | g a.a.[b] | mg a.a.[c] | g a.a.[b] | mg a.a.[c] |
	16g N	g Rice	16g N	g Rice	16g N	g Rice	16g N	g Rice	16g N	g Rice
cyst â→cystine	2.1222	1.42	2.022	1.37	2.101	1.53	1.928	1.24	1.904	1.57
aspartate	8.802	5.89	8.83	5.97	8.825	6.43	8.724	5.62	9.241	7.67
met.sul→met	2.942	1.97	2.919	1.97	2.903	2.11	2.592	1.67	2.449	2.02
threonine	3.464	2.32	3.391	2.30	3.539	2.58	3.353	2.16	3.458	2.85
serine	4.977	3.33	4.872	3.30	5.223	3.81	4.963	3.20	5.097	4.21
glutamate	18.28	12.22	17.84	12.06	19.55	14.25	18.34	11.82	19.06	15.73
proline	4.639	3.10	4.324	2.92	4.774	3.48	4.600	2.97	4.567	3.77
glycine	4.592	3.07	4.524	3.06	4.426	3.22	4.525	2.92	4.546	3.75
alanine	5.795	3.88	5.780	3.91	5.863	4.27	5.849	3.77	6.081	5.02
valine	6.164	4.13	5.870	3.97	6.102	4.45	5.932	3.82	6.097	5.03
isoleucine	4.210	2.82	4.078	2.76	4.301	3.13	4.013	2.59	4.254	3.51
leucine	8.592	5.75	8.436	5.71	8.775	6.40	8.345	5.38	8.708	7.19
tyrosine	4.383	2.93	4.280	2.89	4.416	3.22	4.270	2.75	4.120	3.40
phenylalanine	5.418	3.63	4.980	3.37	5.512	4.02	5.378	3.47	5.646	4.66
lysine	3.675	2.46	3.557	2.40	3.478	2.53	3.492	2.25	3.819	3.15
histidine	2.425	1.62	2.291	1.55	2.531	1.85	2.397	1.55	2.391	1.97
arginine	8.305	5.55	8.232	5.57	8.197	5.97	8.306	5.30	8.610	7.11
proline (440)	4.861	3.25	4.164	2.81	4.779	3.48	4.646	3.00	4.800	3.96
ammonia	2.053	1.38	2.104	1.42	2.153	1.57	2.640	1.70	2.152	1.78
% Nitrogen	1.245		1.234		1.302		1.170		1.486	
% Moisture	14.034		12.323		10.43		11.83		11.129	

a Analyses performed by Dr. Adrian Lamb, Department of Biochemistry, Faculty of Science, Mahidol Univrsity, Bangkok 4, Thailand (personal communication)
b Grams of amino acid per 16 g of nitrogen
c Milligrams of amino acid per gram of field dried rice

irrigated non-glutinous variety, milled by foot-pounding, *hngaw chuek lathin*. The amino acid contents of these varieties lie within the range found in a large series of rice varieties from other parts of Thailand analysed in the same laboratory (Lamb, personal communication). Comparisons between Pa Pae samples must be understood as referring to the single samples described above, with the statistical limitations inherent in this procedure.

Milling Method

Comparison of the foot pounded and machine milled samples of the same rice variety from the same farmer (Table IX) suggests that there is little difference in the proportion of the amino acid to total nitrogen in the oven-dried sample, or in the weight of the amino acid in the field-dried sample. If the result is representative of the two milling methods, the switch to machine milling seems likely to have little direct effect on amino acid content in the diet.

Variety and Soil Type (Swidden Glutinous Rice)

The swidden glutinous variety grown on poor soil is superior to the variety grown on ordinary soil on the basis of the proportion per unit of total nitrogen, and the amount per unit of field dried rice. If the samples are representative of the varieties and of the soil types on which they are normally grown, this implies successful selection by Pa Pae farmers of a superior genetic variety, adapted to poor soil conditions (dry, sandy, low in orgainic content). This further implies the importance of conserving and testing traditional rice varieties for their efficiency in producing nutritionally essential elements.

Non-Glutinous vs. Glutinous Rice

The swidden glutinous variety which is adapted to poor soil is about equal to the non-glutinous swidden variety in amino acid content per unit of nitrogen, but is superior per unit of rice, both due to the higher nitrogen content and lower moisture content of the glutinous variety. Because glutinous rice forms a relatively small portion of the diet for most households, this probably has little nutritional significance.

Swidden vs. Irrigated Rice

Amino acid contents of the swidden non-glutinous foot-pounded sample and the irrigated non-glutinous foot-pounded sample are similar as a proportion of total nitrogen in the samples. The irrigated sample appears to be superior, largely because the total nitrogen content of the irrigated sample is higher, and also because the field-dried moisture content of the irrigated sample is lower. If these samples are representative of swidden and irrigated rice, the results suggest that the switch in emphasis from swidden to irrigated rice production (aside from total productivity) may be nutritionally beneficial.

The conclusions outlined above should be considered as hypotheses to be tested. A full understanding of the variations in amino acid content of Pa Pae rice would require much more elaborate research. This would include analysis of adequate sample size, at least including productivity of different rice varieties per unit of land and per unit of labor; the effects of soil and microenvironment (slope, aspect, temperature, humidity, plant pests, etc.) on productivity of different rice varieties; nutritional content of rice as it is eaten, to allow for differences in moisture content and loss of nutritional substances in preparation; the intake of different varieties of rice (some households, for example, eat very little glutinous rice, while others produce and consume substantial amounts); nutritional deterioration in stored rice.

RICE CONSUMPTION

Pa Pae villagers normally eat three meals per day. Non-glutinous rice is the basis of every meal and by far the major source of calories. Two or three times a week each household spreads out 20 to 40 L of unhusked rice in the sun to dry. Most often it is the older girls and women who then pound it with a foot powered pestle and mortar or take it to be milled at one of the recently introduced power mills. Twice a day, morning and evening, each household boils its rice in a large earthenware or aluminum pot over a wood fire until the water is absorbed. Leftovers from the morning meal are eaten cold for lunch. Farmers take cooked rice, wrapped in a large leaf, to the fields to eat at mid-day.

The most common dish eaten with rice is a stew of a vegetables from the swiddens or the creek-side winter gardens, depending on seasonal availability. Alternatively, villagers may add mushrooms or ferns

TABLE X

Composition of side dishes eaten in Pa Pae with rice (24-hour recall, 20 families, June, 1980)[a]

Adult Meals (60 meals)	No.	Total
Primarily vegetable		37 (61.67%)
With fermented soybeans	29	
With soy beans and/or fish	3	
With no protein	5	
Primarily animal protein		16 (26.67%)
Meat	7	
Meat and vegetables	7	
Fish	1	
Eggs	1	
Primarily pounded peppers and seasoning		7 (11.67%)
Pounded peppers with dried fish	3	
Pounded peppers with soy beans	3	
Pounded peppers with dried fish and soy beans	1	
Extra Meals for Children (9 meals)		
Primarily vegetable		4 (44.44%)
With fermented soy beans	2	
With no protein	2	
Primarily animal protein		5 (55.56%)
Meat	2	
Fish	1	
Eggs	2	

[a] Data from Sally L. Kunstadter (1980), unpublished observations.

gathered near the village as the main component in the single dish eaten whith rice (Table X). The vegetable will usually be cooked with the addition of onions, peppers, salt and a small amount of protein, either dried fish or fermented soy beans (61.67% of adult meals). This vegetable curry may be simmered with water or stir-fried with pork fat. Sometimes, usually when there is heavy work and on all ceremonial occasions, a stew is made with chicken (*Gallus gallus*) pork, (*Sus scrofa*) or both, or more rarely with water buffalo meat (*Bubalus bubalis*) (26.67% of recorded adult meals-possibly a higher than average proportion because the survey was conducted during the rice transplanting season of heavy work and accompanying minor ceremonies). They usually prepare one side dish to eat with rice. If nothing else is available, they make a sauce by pounding chili peppers (*Capsicum frutescens*), salt and onions (*Allium* spp.) with either a small bit of dried fish or fermented soy beans (*Glycine max*) for flavoring (11.67% of surveyed adult meals) (Table IX).

The basic rice and curry meal may be supplemented with a variety of uncooked leafy vegetables or herbs (onions, mint [*Mentha* spp.], coriander [*Coriandrum sativum*], raw bamboo shoots, etc.) and seasonally available uncultivated leaves. When maize (*Zea mays*) begins to ripen in the swiddens in July, it supplements the basic rice and curry diet, as do taro (*Colocasia esculenta*), yams (*Dioscorea* spp.), and manioc (*Manihot utilissima*), when they become available later in the year to add to the basic starch until the next harvest.

A little roasted or preserved meat or salted fish may be served as an extra side dish along with a vegetable main dish at family meals, but few households can afford to do this very often. Children, sick and elderly people are favored with extra animal protein when it is available. Infants are always breast fed, and are usually fed pre-masticated banana and rice gruel and gradually weaned onto solid food. Young children are usually given one or two extra meals of rice, either with curry left over from a main meal (44.44% of recorded children's meals), or with scrambled eggs, dried roasted meat or salted fish, or perhaps some small game such as birds or rats (55.56% of recorded children's meals).

Average rice consumption in February, 1968, about three months after harvest, during the period of heavy work of swidden cutting, was about 710 g *per capita*, suggesting an average daily intake of about 2521 cal from rice (Table XI). If the rice has high protein quality, this implies an adequate amount of both calories and protein from rice alone by WHO and FAO standards.[6] Persons of all ages were included in the measurements from which this average was derived. Rice consumption was highly correlated with body weight ($r^2 = 0.728$). Body weight as a function of age in this population follows an inverted U-shaped curve, with highest weights observed among young adults. Villagers generally agree that young adults are the hardest workers and heaviest producers in the fields. Data from the June,

TABLE XI

Calories in rice consumed by Pa Pae villagers, February, 1968

Mean number of grams of rice consumed person per day 710	Mean number of Calories produced per day per person 3284
Mean number of Calories in rice consumed per person per day 2521	Difference between amount produced and amount consumed per person per day 763

1980 survey, in which rice consumption was estimated by the villagers rather than being weighed, suggest a lower *per capita* average consumption of about 600 g per person per day. The reasons for this difference are unclear. The proportion of very young children in the population was higher in 1980; it was later in the season and families had less rice; a few more fresh vegetables were available; and a larger than normal amount of meat was being consumed. These are all factors which may have lowered the rate of rice consumption, aside from the differences in measurement methods. These examples illustrate some of the difficulties in making estimates of average and total consumption.

According to our figures, villagers produced a daily *per capita* surplus of about 763 cal beyond the number of rice calories eaten in 1967 (Table XI). We do not have precise information on the disposition of this apparent surplus. Some was lost to rodents, insects and other pests; some was fed to domestic animals, some consumed (primarily by older men) in the form of rice liquor; some was traded or sold; but the apparent surplus was misleading. In fact the village was in a serious rice deficit following the 1967 harvest because of indebtedness. Pa Pae villagers paid out a total of 56,120 l of rice, 31% of their production, including 34,580 l (19% of production) to repay debts and mortgage interest between the 1967 and 1968 harvests. The net effect was that only four households (8%) reported they had enough of their own rice to last until the 1968 harvest. The remaining households had to borrow, buy or do without.

We do not have comparable information for the 1980 season, but the economic situation in the village now seems better. One measure of this is the amount and proportion of early rice grown. Early rice is a riskier crop than late rice because it is harvested before the end of the rains, and it does not yield as well as late rice. Farmers plant it in the hope that they can harvest some a month or two before the regular harvest. Thereby they can stave off starvation when their rice supplies are exhausted and when the price of rice (if it can be found) and effective interest rates are at their highest in the hills. In 1979 early rice amounted to 3.2% of the total planted as compared with 6.5% of the total in 1967.

Another measure of the improved economic position of the villagers is the number of irrigated rice fields owned and the proportion mortgaged. Mortgaging was traditionally a sign of economic desperation or imprudence, since the high interest rates (50% per annum, more or less) generally precluded repayment. In June 1967, 8 of 26

TABLE XII

Mortgages on Pa Pae irrigated fields, June, 1967, and June, 1980

	June 1967	June 1980
Total number of irrigated fields	26	55
Number of fields mortgaged (net)[a]	8	8
Percent of fields mortaged (net)[a]	30.8%	7.3%
Amount owed (approx. 1967 baht, 20 baht = $1.00 U.S.)[b]	15,500	6,000

[a] In 1980 five fields were mortaged, but one was in effect an exchange. In this transaction the Pa Pae villager owed 10,000 baht on the field he had mortgaged and was owed 11,000 baht on the field mortaged to him.

[b] Net mortgage indebtedness of 12,000 baht in 1980 was converted to 1967 baht by estimating a 50% decline in the value of the baht since 1967. The actual rate may have been higher than this. The annual rate of inflation for Thailand as a whole was estimated at over 16% during the first half of 1980, associated with the drastic increases in costs of imported fuel. It is difficult to calculate the inflation rate at Pa Pae, which has a predominantly subsistence economy. Rice prices have increased a total of about 50%, animal prices about 200%, and price of silver (the most important traditional store of value) at world rates of about 800% in the 13 years under study.

fields (30.8%) owned by Pa Pae villagers were mortgaged for 15,500 *baht*, while in June 1980 only four of the 55 irrigated fields (7.3%) were mortgaged for an equivalent of a net of about 6,000 baht (in 1967 baht) (Table XII). The villagers' recent success in paying off mortgages or in avoiding them seems related to three factors: income from wage work in the village associated with government development programs, availability of local credit at relatively low cost, and possibly inflation. The first of the local credit facilities was a Royally sponsored rice bank, established early in 1970. Late in the 1970s Catholic villagers banded together to form their own rice bank. These banks charge only enough interest to make up losses in operation, not to make a profit. Ironically, availability of wage work as a source of cash, and availability of low cost credit, plus inflation (an increase of wages relative to the price of rice), which makes repayment of old debts easier, seem to have been much more effective in reducing the problem of indebtedness than have the cash cropping projects (silk worms, soy beans, coffee) which have been the main thrust of government development efforts over the past decade.

The subsistence rice economy of Pa Pae villagers now seems stronger than in the late 1960s. A slightly higher proportion of households was working fields of their own (96.4% in 1979, 94.6% in 1980) as compared with 1968 (92.7%), and a much higher proportion of

c

TABLE XIII

Number of Pa Pae households engaged in swidden and irrigated rice farming, 1968, 1979 and 1980

	1968	1979	1980
Households with swiddens	44 (86.3%)	45 (81.8%)	45 (80.4%)
Households with irrigated fields	26 (51.0%)	44 (80.0%)	45 (80.1%)
Households with both swidden and irrigated fields	23 (45.1%)	36 (65.5%)	37 (66.1%)
Households with swidden only	21 (41.2%)	9 (16.4%)	8 (14.3%)
Households with irrigated fields only	3 (5.9%)	8 (14.5%)	5 (8.9%)
Households with neither swidden nor irrigated fields	4 (7.8%)	2 (3.6%)	3 (5.4%)
Total Households	51	55	56

households now has irrigated fields (80% in 1980 vs. 51% in 1968) (Table XIII). The future of the subsistence economy is increasingly threatened, however, by population growth at Pa Pae and in neighboring villages. This leads to land shortages and possibly to irrigation water shortages. At the same time there is increasing competition for land for other purposes, especially for government reforestation and development projects of little direct benefit to the villagers. Better access to Pa Pae as a result of the construction of a dry season road in 1979 has also increased pressure on local forest resources. People from all around Pa Pae have come to cut the aromatic bark from trees in fallow swiddens which they sell at the road head. The villagers sell the bark for a small amount, which probably does not compensate for the damage to forest regeneration and restoration of soil fertility in the fallow swiddens.

CONCLUSION: REVIVAL OF THE SUBSISTENCE ECONOMY THROUGH AGRICULTURAL INTENSIFICATION AND INCREASED ACCESS TO CASH AND CREDIT

The subsistence economy of the Lua' hill villagers of Pa Pae was under considerable stress in the late 1960s. Although production from swiddening and irrigated rice farming appeared to be sufficient for subsistence needs, the increasing burden of subsistence indebtedness and population growth meant that 92% of village households did not have enough rice to last from one harvest to the next. In 1980 the amount of indebtedness, as indicated by numbers and amounts of mortgages on irrigated fields, and the proportion of early rice grown seemed much reduced. Despite the continued rapid increase in population and the associated increase in dependency of the population on a smaller proportion of workers, the economic situation appeared to be better. Productivity per worker had increased greatly, as emphasis had shifted from swidden to irrigated rice. Reduced indebtedness seems to have come about as a result of increased availability of wage labor, low cost credit within the village and, incidentally, inflation, which has eased the repayment of old debts, rather than as a result of modernization of agriculture or development of cash crops.

At the same time, the authority of village leaders and the cohesion of the village as a land managing unit seems diminished. This is a result of external changes (growth of population and expansion of nearby villages, expansion of administrative authority of the central government without confirmation of traditional village land boundaries, introduction of other authority figures such as teachers into the village, introduction of Christianity, etc). Meanwhile, swiddening has lost its economic centrality and village leaders have no role in coordinating irrigated agriculture.

An unresolved historical question remains: Why did not the Pa Pae villagers turn to irrigation long ago, since it allows production of more rice with less effort once the initial investment has been made in levelling fields and digging ditches? Wet rice technology used in Pa Pae in 1980 is traditional. No chemical fertilizers or pesticides are used, nor is "miracle rice" grown, and no fossil fuel powered equipment is being used in rice production. Wet rice farming began at Pa Pae in the early 1930s and was being practiced in some nearby hill Lua' villages at least since the 1880s (Hallett 1890). The Lua' of Pa Pae have never been isolated from the lowland, wet rice growing Northern Thais. Lack of capital was not so much of a problem in the past, when Pa Pae villagers were wealthier and received rent (in the form of 10% of the annual rice crop) from Karens who begged permission to settle on Lua' land.

Perhaps the failure to expand irrigation was a result of the vested interest of religious leaders in maintaining the traditional system in which they controlled swiddening operations and access to swidden land. Perhaps the motivation for intensifying agriculture was weak until the villagers saw (as most of them clearly do now) that population pressure was threatening their traditional swidden economic system. Although cause and effect cannot be easily demonstrated, the increased emphasis on irrigated agriculture and the increased productivity per worker has occured simultaneously with the weakening of the authority of traditional village leaders and the increased individualization of control over land. It is too soon to tell whether the renewed success of the subsistence economy has been gained at the expense of disintegration of the village.

NOTES

1. Revised version of paper prepared for presentation in Symposium on Food Energy in Tropical Ecosystems, American Anthropological Association Annual Convention, Washington, D.C., December 6, 1980.
2. Research has been supported in the past by National Institute of General Medical Sciences, National Geographic Society, National Science Foundation, Princeton University, University of Washington and the East-West Center. Current work is supported by Grant BNS-7914093 from the National Science Foundation and by the Population Institute, Resource Systems Institute and Environment and Policy Institute, of the East-West Center. The assistance of support staff at the East-West Population Institute, especially Ms. Ruby Bussen and Ms. Gayle Uechi of the data processing section and Ms. Mary Kehoe of the secretarial staff, is gratefully acknowledged, as is the help of the Research Institute for Health' Sciences, Chiang Mai University, especially Ms. Suvipa Khumboonruang and Ms. Kayoon Panas-Amphol. The research would not have been possible without the continued generous cooperation of the Pa Pae villagers and the sponsorship of the National Research Council of Thailand.
3. A perceptive and useful description of northwestern Thailand in the late 19th century is to be found in Hallett (1890). A discussion of feudal relations between Lua' villages and Northern Thai princes appears in Nimmanahaeminda (1965). Detailed description of Pa Pae in the 1960s appears in Kunstadter (1966a, 1966b, 1967, 1969, 1971). Ecological aspects of the swidden cultivation system at Pa Pae are described in Kunstadter (1978a, 1978b); Kunstadter, Sabhasri and Smitinand (1978); Sabhasri (1978); and Zinke, Sabhasri and Kunstadter (1978). For a discussion of the amount of work peasants will do, see Chayanov (1966).
4. In 1980, because of land shortage, Pa Pae villagers cut swidden blocks in two separate places. Farmers working in one area, fearing an early start of the monsoon rains would dampen the fuel bed, started their swidden fire without permission of the village leaders. This was on a Sunday and the Christians refused to patrol to prevent the fire from escaping. A hut in the irrigated field of a farmer from a neighboring village was burned because he had not been notified in time to remove the highly flammable roofing from the shelter. This farmer demanded the payment of damages, but after several months of wrangling, the Pa Pae headman was unable to persuade his villagers to accept responsibility for the damage. Such lack of coordination and discipline was unheard of a decade earlier.
5. If a household has irrigated fields and has only one working age male, it may hire a young boy from a poor family to lead the water buffalo during the time of field preparation.
6. Rice samples were analyzed by Institute of Food Research and Product Development, Kasetsart University (personal communication 1981).
7. World Health Organization (1974).

REFERENCES CITED

Chayanov, Aleksander V. 1966. *Theory of Peasant Economy*. Homewood, Illinois: Irwin.

Hallett, Holt S. 1890. *A Thousand Miles on an Elephant in the Shan States*. Edinburgh and London: Wm. Blackwood.

Kunstadter, Peter 1966a. Irrigation and social structure: narrow valleys and individual enterprise. Paper presented at the Pacific Science Congress, Tokyo, 1966. Abstract published in Proceedings of the Congress.

_____ 1966b. Residential and social organization of the Lawa of northern Thailand. *Southwestern Journal of Anthropology* 22(1):61–83.

_____ 1967. The Lua' and Skaw Karen of Maehongson Province, northwestern Thailand. pp. 639–674 *in Southeast Asian Tribes, Minorities and Nations*, ed. by Peter Kunstadter, Princeton, N.J.: Princeton University Press,

_____ 1969. Socio-cultural change among upland peoples of Thailand: Lua' and Karen— two modes of adaptation. *Proceedings of the Eighth International. Congress of Anthropological and Ethnological Sciences* (1968) 2:232–235. Tokyo: Science Council of Japan.

_____ 1971. Natality, mortality and migration of upland and lowland populations in northwestern Thailand. pp. 46–60 *in Culture and Population*, ed. by Steven Polgar, Carolina Population Center Monograph. Chapel Hill, N.C., and Cambridge, Mass: Schenkman

_____ 1978a. Subsistence agricultural economies of Lua' and Karen hill farmers, Mae Sariang District, Mae Hongson Province, northwestern Thailand. pp. 74–133 *in Farmers in the Forest*. ed. by Peter Kunstadter, E. C. Chapman and Sanga Sabhasri. Honolulu: University Press of Hawaii,

_____ 1978b. Ecological modification and adaptation: an ethnobotanical view of Lua' swiddeners in northwestern Thailand. pp. 168–200 *in The Nature and Status of Ethnobotany*. ed. by Richard I. Ford. Museum of Anthropology, Anthropological Papers 67. Ann Arbor, Mich: Univ. of Michigan

Kunstadter, Peter, Sanga Sabhasri, and Tem Smitinand. 1978. Flora of a forest fallow farming environment in northwestern Thailand. *Journal of the National Research Council of Thailand* 10(1):1–45.

Nimmanhaeminda, Kraisri. 1965. An inscribed silver-plate grant to the Lawa of Boh Luang. pp. 233–238 in *Felicitation Volumes in Southeast Asia Studies Presented to His Highness Prince Dhanivat Krmamum Bidyalabh Bridyakorn*, vol. 2. Bangkok: The Siam Society.

Sabhasri, Sanga. 1978. Effects of forest fallow cultivation on forest production and soil. pp. 160–184 *in Farmers in the Forest*, ed. by Peter Kunstadter, E. C. Chapman and Sanga Sabhasri. Honolulu: University Press of Hawaii.

United States Department of Agriculture. 1963. *Handbook 8*. Washington, D.C.: U.S. Government Printing Office.

World Health Organization. 1974. Handbook on Human Nutritional Requirements. World Health organization Monograph Series, no. 61. Geneva: World Health Organization.

Zinke, Paul, Sanga Sabhasri, and Peter Kunstadter. 1978. Soil fertility aspects of the Lua' forest fallow system of shifting cultivation. pp. 134–159 *in Farmers in the Forest*, ed. by Peter Kunstadter, E. C. Chapman and Sanga Sabhasri. Honolulu: University Press of Hawaii

LIST OF SCIENTIFIC NAMES OF FOODS
(in the order in which they appear in the text)

Rice, non-glutinous	*Oryza sativa*
glutinous	*O. sativa* var. *glutinosa*, or *O. glutinosa*, or *O. japonica*
Chicken	*Gallus domesticus*
Pork (pig)	*Sus scrofa*
Water buffalo	*Bubalus bubalis*
Rat	*Rattus* spp.
Chile pepper	*Capsicum frutescens*
Onion	*Allium* spp.
Soy bean	*Glycine max*
Mint	*Mentha* spp.
Coriander	*Coriandrum sativum*
Bamboo	several genera
Maize	*Zea mays*
Taro	*Colocasia esculenta*
Yam	*Dioscorea* spp.
Manioc	*Manihot esculenta*
Banana	*Musa* spp.

PART II
SUBSISTENCE STRATEGY

CHAPTER 3

SOCIAL IMPLICATIONS OF DIETARY PATTERNS IN THREE COMMUNITIES OF AMAZONIAN PERU

ELOIS ANN BERLIN
Division of Family Medicine
Stanford School of Medicine
Stanford University
Stanford, California 94305

ABSTRACT

The environmental and geomorphological context, with its distinctive floral and faunal biomass, delimits the potential resources which can be utilized by human populations. No human group ever utilizes all of the potential nutrient sources in their environment. Which natural resources are translated into dietary resources are cultural decisions.

Human populations have exhibited the unique tendency to modify the natural resource base through the development of agriculture and by the domestication of animals. These are cultural interventions directed at the establishment of a reliable resource base for the most fundamental of human needs, food consumption.

Food production is constrained by ecological setting but food preferences, determined by culture, dictate exploitation strategy in terms of hunting, fishing, gathering, crop production and animal husbandry. Because assurance of adequate nutrition is such a basic human need, it takes on great social significance. Sharing of food is a caring gesture. Withholding of food is a hostile action. When social groups of different traditions come together, dietary practices may be modified because of culture change or competition for resources.

The ecological and cultural variation represented by the three communities in this study provide an opportunity for observation of the environmental, cultural and socio-political influences on dietary patterns and of the forces of change.

INTRODUCTION

Numerous factors influence dietary patterns. Food resources are determined to a major degree by environmental setting and ecological conditions (Loomis 1976). However, *which* potential edible resources become defined as foods are cultural decisions (Lee 1957; Sauer 1954). Further, practices that modify the resource base, such as agriculture and animal husbandry, while constrained by environment, are fundamentally cultural interventions (Boserup 1965). Additionally, social structure (Greene 1977; Montgomery 1977; Seagraves 1977) and

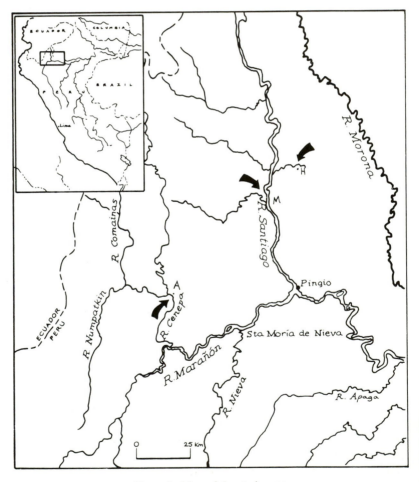

Figure 1. Map of the study area

social relations (Briggs and Calloway 1979:373–388) have frequently been shown to affect food distribution within households (Gross and Underwood 1972) and within and between social groups (Cassel 1957; DeWalt, Kelly and Pelto 1980). (See also Gifft, Washbon and Harrison 1972; Fitzgerald 1977; Jerome, Kandel and Pelto 1980.)

The following discussion will focus on the significance of such factors on diet in three communities of the Peruvian Amazon (see Figure 1.) The research reported here[1] was carried out on two different major tributaries of the upper Marañon River in north central Peru, the Cenepa and the Santiago. Each of these rivers exhibits slightly different ecological characteristics. These differences allow for a comparison of environmental influences on two rather closely related Jivaroan communities, one Aguaruna (Fig. 1, A), and the other Huambisa (Fig. 1, H). The third community (Fig. 1, M) is comprised of Peruvian migrants, most of whom settled in the region since 1970. This community allows for elucidation of the influence of differing cultural patterns on diet since the migrants share the same environment as the Huambisa.

The juxtaposition of the migrants with territories previously controlled by Indians presents a situation that yields insights into the influence of between-group social relations on the utilization of diet and food resources. The heterogeneous migrant community demonstrates patterns that can be interpreted as indicators of perceptions of social prestige and upward mobility. Cultural changes may also be reflected in resource utilization and, in the Aguaruna case, extend even to a triggering of the breakdown of food taboos which have a strong rationale in native mythology.

Finally, the above factors, in conjunction with the success of the subsistence strategies in each of the three groups, allows for an evaluation of resource management and future trends in adaptive change, at least at current population levels.

BACKGROUND

The approximate distribution of the Aguaruna and Huambisa is defined broadly by the course of the Upper Marañon River and its tributaries in the Department of Amazonas, Peru. The Aguaruna number some 20,000 and the Huambisa range to about 5000 individuals (Guallart 1964; Uriarte 1976). The Aguaruna community chosen for this study lay on the Cenepa River. This is an area of

Figure 2. Aguaruna children canoeing down narrow stream, typical of Site 1

tropical rain forest characterized by abrupt, rugged terrain cut by
numerous, relatively fast-moving creeks and streams. The elevation is
approximately 700 feet (213 m) but nearby ridges rise quickly to
1500–2500 feet (450–760 m). Although there are periods of greater
and lesser rainfall, there is no marked dry season. Temperatures range
around 80° F (26° C) (Figure 2).

To the east lies the Santiago, a river considerably larger than the
Cenepa. It empties into the Marañon just above the Pongo de
Manseriche, the boundary marker between Peru's *montañá* region and
the true Amazon Basin. In contrast to the steep hills and rapid waters
of the Cenepa, the Santiago Valley, at approximately 600 feet (180 m)
elevation, is characterized by open, sweeping vistas and, in places,
broad sandy beaches. *Mauritia* palm forests are typical in areas of
annual flooding. There appears to be somewhat more marked
seasonal contrast in rainfall on the Santiago. Ox bow lakes (*cochas*)
are common and creeks flowing into the Saniago are slow and
meandering. The Huambisa community lay on one of these small
meandering creeks about four hours by canoe from the main river
(Figure 3).

Finally, the third community of Peruvian migrants was situated
directly on the Santiago proper. Its state-owned lands lie on the

Figure 3. Vista from inland Huambisa Village, Site 2

Figure 4. Merchant's boat moored at village landing, Site 3

boundary between Aguaruna and Huambisa territory. These migrants settled in their present location in the early 1970s as part of the Peruvian government's colonization program for developing Peru's tropical forest areas (Figure 4).

METHODS

Nutritional data were collected in each of the three communities. The data base represents over 15,000 person days of weighed dietary intake as seen in Table I.

The data were recorded as intake per household for varying periods of time. In order to make resuls for the three groups comparable, the total summed intake for each class of foods was divided by the number of person days represented by each community study. Since household composition varied, an attempt was made to normalize the number of people for calculation of *per capita* intake. Small children were assigned a value of 0.5, school-aged children were assigned a value of 0.75 and adults a value of 1.0. In this way, it,was hoped to avoid an apparent decrease in intake which would actually be a reflection of larger numbers of children in a study group. The *normalized person*, then, is an attempt to make the food consumption reflect the food needs of the population. The values assigned were somewhat arbitrary, but logical.

TABLE I

The Study Populations and Data Base

Group	Person Days	Normalized No. Persons	No. Days	No. Families
Aguaruna	5502	21.00	262	5
Huambisa	3005	29.75	101	6
Colonist	7395	51.00	145	10
Total	15902	101.75	508	21

DIET

Table II shows the *per capita* bulk consumption of foods eaten in the three communities. It was calculated by dividing the total bulk

TABLE II

Daily bulk *per capita* food consumption of three Peruvian groups (gms)

	Aguaruna	Huambisa	Colonist
Banana	414.5	336.0	627.0
Fruit	30.0	108.0	11.0
Grains	12.0	3.0	231.0
Greens	1.5	— —	1.1
Manioc	1045.0	822.0	154.0
Nuts & Beans	7.0	— —	18.7
Palm	14.5	84.0	3.3
Tuber	64.0	42.0	22.0
Bird	35.0	35.4	56.1
Fish	100.0	120.0	69.3
Reptiles & Amphibians	1.5	1.2	16.5
Mammal	45.0	67.8	90.2

consumption for each community by the number of person days represented.

Bulk food consumption is an indicator of both food preferences and availability. Manioc is the major staple for the Aguaruna and Huambisa (Figure 5). It is native to the region and is the primary cultivar for many Amazonian Indian groups. Bananas represent a major contribution to diet for all three groups. It is the second major staple for both Indian groups and the primary staple, by weight, for the colonists (Figure 6).

The second staple food class for the migrants is grains. When the grains consumed by the colonists are adjusted to account for their measurement as dry weight in Table II, the consumption figure rises to 517 g. *per capita*. This is a more realistic figure and is borne out by the frequency data to be presented below. Grains contribute relatively little to Indian diets, even though they have begun growing some on

Figure 5. Aguaruna woman working in manioc garden

the Santiago for export (Figure 7).

Miscellaneous tubers (all root crops except manioc) rank second as a carbohydrate source for the Aguaruna but are surpassed by palm heart for the Huambisa. This is probably a reflection of resource availability. The palm forests of the Santiago do not occur on the Cenepa. Both heart of palm and palm fruits are highly valued foods.

Fish is the primary source of protein for the Indians of both rivers. Although it is a major protein source for the migrants, fish is secondary to mammals in their diet. Except for infrequent fishing expeditions employing the native system of poisoning waters, fishing is a rather casual endeavor for the colonists, with younger family

Figure 6. Huambisa woman's new manioc garden with banana trees in foreground

Figure 7. Colonist male sitting on tree stump in new rice field

members spending more time fishing than adults. This probably accounts for the lower fish quantities consumed by the colonists.

Mammal sources provide roughly half the amount of dietary protein that fish do for the Aguaruna and the Huambisa. The relative contribution of fish and mammals to the Indian diet must be viewed as a reflection of resource availability, because they value game animals more highly than any other food.

The high proportion of reptiles in the colonists' diet is primarily from river turtles. These turtles are found on the sandy beaches along the Santiago, a condition absent on the Cenepa. The Huambisa community is not located on the Santiago proper, but lies, as mentioned earlier, on a small stream. On occasion, no doubt, they harvest some turtles from the Santiago, but none were included in this four month study period.

Another measure of food preferences and dietary patterns is

TABLE III

Frequency of consumption of selected foods by three Peruvian groups

	Aguaruna	Huambisa	Colonist
Banana	1.7xday	0.9xday	1.9xday
Fruit	0.3	0.16	0.2
Grains	0.2	0.01	1.5xday
Greens	0.02	— —	0.02
Manioc	2.1xday	0.8	0.6
Nuts & Beans	0.05	— —	0.2
Palm	0.09	0.17	0.01
Tuber	0.2	0.1	0.12
Bird	0.4	0.13	0.3
Fish	0.9	0.4	0.5
Reptiles & Amphibians	0.03	0.008	0.1
Mammal	0.4	0.15	0.7

frequency of consumption. Table III shows the frequency of food consumption calculated for the three communities and was derived by dividing the total number of times a food was reported as eaten by the total number of days represented by all households in each community. For the most part, the frequency results parallel the quantity figures.

The data sets are not equivalent in all respects and there is some distortion of frequency figures for the Huambisa. The Aguaruna families represented here were literate and, therefore, were instructed to write down each time a food was consumed but to report a weight only at the time of preparation. The Huambisa were non-literate recorders who marked lines on pictures of scales to indicate food weights at the time of preparation. If portions of the same food were consumed at subsequent meals (for example, a large game mammal smoked and eaten over a period of days), no note was made of it after the initial weighing. Hence, frequency of food consumption is under-reported for the Huambisa. Since seven families from the community of migrants were also non-literate, under-reporting constitutes a slight margin of error in these data as well. (See Berlin, 1981:42–48 for an

TABLE IV

Daily per capita bulk and frequency of consumption of foods by three Peruvian groups translated into serving portion and time interval

Food Class	Aguaruna	Huambisa	Colonists
Fish	$\dfrac{226 \text{ g}}{1.1 \text{ days}}$	$\dfrac{510 \text{ g}}{2.5 \text{ days}}$	$\dfrac{126 \text{ g}}{2 \text{ days}}$
Mammal	$\dfrac{230 \text{ g}}{2.5 \text{ days}}$	$\dfrac{731 \text{ g}}{7 \text{ days}}$	$\dfrac{123 \text{ g}}{1.5 \text{ days}}$
Bird	$\dfrac{183 \text{ g}}{2.5 \text{ days}}$	$\dfrac{472 \text{ g}}{8 \text{ days}}$	$\dfrac{175 \text{ g}}{3.3 \text{ days}}$
Bananas	$\dfrac{488 \text{ g}}{1.7 \times \text{ day}}$	$\dfrac{507 \text{ g}}{1.1 \text{ days}}$	$\dfrac{330 \text{ g}}{1.9 \times \text{ day}}$
Manioc	$\dfrac{1055 \text{ g}}{2.1 \times \text{ day}}$	$\dfrac{2055 \text{ g}}{1.25 \text{ days}}$	$\dfrac{262 \text{ g}}{1.7 \text{ days}}$
Grains	$\dfrac{100 \text{ g}}{5 \text{ days}}$	$\dfrac{500 \text{ g}}{100 \text{ days}}$	$\dfrac{154 \text{ g}}{1.5 \times \text{ day}}$
Tuber	$\dfrac{490 \text{ g}}{5 \text{ days}}$	$\dfrac{660 \text{ g}}{10 \text{ days}}$	$\dfrac{176 \text{ g}}{8 \text{ days}}$

in-depth discussion of dietary collection methods with both literate and non-literate subjects.)

The data from Tables II and III can further be used to calculate serving size per time interval in days, as shown in Table IV. This is a more realistic representation of dietary patterns since it reflects the approximate portion served and the time interval between servings of a given food. Examination of selected foods reveals that average meal portions for both Indian groups is larger than serving size for the colonists.

Comparison of the two Indian data sets allows for some estimates of quantity (by weight) of fish capture. For all Huambisa households reporting, a combined catch weighing approximately 15 kg (510 gm × 29.75 persons = 15,173 gm) (33 lb) is taken on the average of about three times per week (every 2.5 days). Mammals were taken on average once a week, and the average take was 21.8 kg (731 gm × 29.75 persons) (48 lb). These figures, compared with the smaller portions and more frequent servings reported by the Aguaruna, might imply how long an average catch lasted. This inference does not, however, take into account ecological differences which could influence size and frequency of capture.

An underlying assumption of the preceding discussion has been that cultural preferences explain the major differences in diet between Indian and colonist populations. As mentioned earlier, the colonist community is not a homogeneous population. In four households, both husband and wife came from the highlands of northern Peru. Two couples had been long-term residents of the immediate region of the greater Marañon Valley, migrating in from other mestizo[2] towns. One couple represented second generation mestizos from the region. Three people came from the Iquitos area of eastern Peru. One of the women in this group was of Peba Indian descent and brought with her bitter manioc processing techniques. Seven families were made up of an Aguaruna or Huambisa female and a non-Indian male. These men had been long-term residents of the area and their families did not represent part of the early 1970s colonist influx. They also tended to exhibit subsistence strategies closer to that of the Indians, with the women planting traditional gardens.

The colonist males who farmed planted seed crops (rice and corn) for family consumption and for cash export. The females who gardened produced root crops, beans, fruits, vegetables and herbs for family consumption, including exchange with neighbors. Potatoes, which will not grow in this environment, had probably been the primary staple of the highland migrants in their homelands and it is of interest to note that bananas, not manioc or other root crops, play such a major role in their diet. This probably has to do with ecological and social factors. Bananas are among the easiest of cultivars. They command premium prices in urban centers—and thus are prestige foods.

The emphasis on rice probably has economic as well as social implications. First, there is a high market demand for rice. All surplus produced is marketable if transport problems can be solved. Secondly, rice is the staple of the upper classes in Peru. Migrants are usually people who seek some perceived betterment of their quality of life. Higher socio-economic status, with concomitant improvement in nutrition and health, are common expectations. Behavioral changes reflect these expectations and consumption of prestige foods symbolizes upward mobility. The high quantity of rice in the colonists' diet is an example of these attitudes and aspirations.

Those colonist males not primarily involved in planting seed crops were usually involved in other professions. Some were merchants, others served as employees of the school system, and others periodically panned for gold. Some combined farming with their other livelihood activities. In any case, their time distribution and subsistence orientation was not toward hunting and fishing as a major activity. Animal food tended to come from domestic animals. Animal husbandry was primarily in the hands of women. The relative contributions of domestic versus game mammals and birds seen in Table V reflect these cultural patterns. While the colonists have

TABLE V

Relative consumption of game and domestic animals by three Peruvian groups

| | Aguaruna | | Huambisa | | Colonist | |
	Freq. Percent	Percent Total Weight	Freq. Percent	Percent Total Weight	Freq. Percent	Percent Total Weight
BIRDS						
Native	43	23	81	73	3	3
Domestic	57	77	19	27	97	97
MAMMALS						
Native	92	91	100	100	26	27
Domestic	8	9	0	0	74	73

adopted a subsistence strategy based primarily on introduced crops and domesticated animals, they have achieved a high degree of autonomy. Grain crops represent about 80% of their dietary staples. Rice makes up 67% of the grains consumed and corn another 11%. Both are raised by the colonists. The remaining grains, primarily wheat products (18%) and some oats, are imported. Until a rice mill was brought into the village, the common practice was to export the unmilled rice and purchase the refined product. Now local residents can sell rice to the mill for cash and, for a fee, mill their own rice for personal consumption. With this step, they have gained greater self-reliance and are less dependent on the national market system.

The indigenous populations have developed an autonomous subsistence system based on swidden agriculture supplemented by hunting, fishing and gathering. Aguaruna and Humabisa women are the primary cultivators. The Indian males' agricultural role is limited to clearing new fields for manioc gardens and the cultivation of seed crops such as native peanuts, corn and introduced rice.

Hunting has always been the responsibility of Jivaro males. Fishing is undertaken by males and/or females, depending on the technique to be used. Historically, animal husbandry was limited to palm grub production, possibly the native muscovy duck, and occasional wild animals raised as pets, but frequently ending up in the soup pot. The Indians' food resources are still those of tradition and their interaction with the national market system is quite recent. There are, however, signals of culture change in the food resources, especially with the domestic animals of the Aguaruna.

One distinct signal of culture change is the very high percentage of domestic birds, primarily chickens, found in the Aguaruna diet: 77% by weight and 57% by frequency. The difference in the percentages by weight and frequency is attributable to the larger size of chickens and ducks contrasted with wild birds, which can be very small. The Huambisa have also begun to raise chickens and ducks but their contribution to diet is much smaller than for the Aguaruna. Domestic mammals, primarily pigs, have also begun to contribute to the Aguaruna diet. By far the more interesting aspect of mammal consumption, however, comes from examination of the species of game mammals consumed.

The Aguaruna, like many Amazonian groups, have traditionally tabooed consumption of certain animals. (See Table VI.) These taboos are beginning to break down. The dietary records show that 14% by frequency and 15% by bulk of all wild mammals consumed by the Aguaruna were the formerly tabooed deer and capybara.

TABLE VI

Changes in game mammal utilization by Peruvian Aguaruna and Huambisa

	Aguaruna		Huambisa	
	Freq. Percent	Percent Total Weight	Freq. Percent	Percent Total Weight
Tabooed	14	15	0.8	0.8
Game	86	85	99.2	99.2

One myth states that deer are not eaten because they may be temporarily visible human souls which wander in the forest (Harner 1972:150). Another explains the inedibility of tabooed animals as a result of contamination (Chumap and Garcia-Renueles 1979:103). As potential human souls, the deer would possibly be under a stronger taboo than the animals made inedible by contamination. In fact, however, there were 65 reports of deer consumption versus one report of capybara being eaten by the Aguaruna. The Huambisa reported only one instance of a tabooed animal, a deer, having been eaten. Additionally, the Aguaruna report five instances of caiman consumption, comprising 12% of all reptiles and amphibians eaten. Caiman was among the formerly tabooed animals.

The difference between the two Indian groups' consumption of tabooed animals is more striking when it is realized that the Aguaruna data are reports from 1975. Huambisa data were recorded in 1979–80. Changing food patterns among the Aguaruna, contrasted with the Huambisa, are very likely due to demographic and environmental variables. Villages on the Cenepa, although usually smaller than those on the Santiago, are much more closely spaced, giving the impression of almost continuous habitation. In contrast, one travels long distances between settlements on the Santiago.

Traditionally, the Santiago region has always been characterized by abundant game. The Aguaruna also report that game used to be much more plentiful on the Cenepa. These dietary data support both statements. The inference can be made that changing settlement patterns from small, isolated hamlets to concentration in virtually contiguous villages along the river has led to decreased local game supplies, probably due to over hunting and the movement of game away from settled areas. This, in turn, has resulted in a redefinition of food resources to include tabooed animals and contributes ultimately to the greater reliance on animal husbandry, particularly domestic fowl.

Intergroup relations between the Indians and the colonists are ambivalent. On the one hand, Indians view the colonists as resources. They buy manufactured goods from the merchants and sell rice, game, fish and produce to the migrants. The colonists also employ some occasional Indian laborers.

Five colonists (3 male, 2 female) were employees of the Indian school system and had close relationships in the Indian communities. Seven migrants were married, as noted earlier, to Indian women, five of whom had relatives in nearby Indian communities. Just as the Indians purchase patent medicines, colonists occasionally seek treatment from Indian curers. There are many avenues of social interchange and intergroup cooperation. There is, however, a concurrent underlying tension. The colonists are viewed as intruders who are usurping Indian lands and the right of the government to grant these lands to outsiders is questioned.[3]

These social relations are reflected in diet. The migrants, because of the political constraints outlined, are reticent to exploit resources on lands not specifically consigned to them by the government. They receive strong encouragement in this attitude from the Indians. They live, for example, in the region with expansive palm forests characteristic of the broad, relatively unbroken terrain of the Santiago Valley. The low consumption of palm by the colonists contrasted with the large quantities consumed by the Huambisa of the same region (Tables II, III) is at least in part a result of these social constraints. The same is true of other forest products, though perhaps to a lesser extent for game. The river is not defined territorially in the same way that land is and river resources play a major role in the colonists' diet.

There is an economic constraint on inclusion of game birds in all three diets. The price of shotgun shells is currently so high as to prohibit their purchase for bird hunting. The native blow-gun is ideal for killing birds and arboreal mammals, and this part of the traditional male role is reflected in the larger amounts of game birds eaten by Indians. This difference is even more dramatically evident in the species counts of birds consumed (not shown here).

SUMMARY AND CONCLUSIONS

An examination of the diets of an Aguaruna and a Huambisa community residing in slightly different ecological settings and that of a group of migrants who shared the environment of the Huambisa reveals contrasting dietary patterns and allows for an analysis of factors effecting those patterns. Social and cultural preferences are the primary determinants of food selection, given the limitations of the environment. These include traditonal food patterns, social values of foods, and intergroup social relations. Demographic variables and culture change also appear to play a role in dietary patterns.

Either agricultural strategy, that of the Indians or that of the migrants, are viable alternatives for local food production in the communities studied. Both manioc cultivation and seed crop cultivation can meet local needs and produce a surplus. However, although manioc is a cultivar which produces easily and abundantly, the fresh tubers must be consumed within a few days after harvesting. Any surplus production cannot be moved to market without processing. The technology and equipment for processing it to tapioca, fariña or animal feed, for example, do not exist in this area. This fact, in conjunction with the implications of the demographic patterns on the Cenepa and the direction of culture change on both rivers, implies that, in the long term, seed crop cultivation and reliance on domestic animals will increasingly be the pattern for both Indian groups. The trend for the future would seem to be a closer approximation of the economic and subsistence strategies of the migrants as the Indians move toward cash economies and participation in the national market system. Care must be taken, and future research should be directed toward evaluation of the environmental implications of this trend.

NOTES

1. The research reported here was supported by NIMH Grant 22012 and NSF Grant BNS76-17485.
2. The term "mestizo" is used in this region to designate settlers of mixed Indian-non-Indian heritage who adopt the lifestyle of the national culture. It is a term frequently used for self-reference and does not carry negative connotations. Persons of mixed heritage who adopt Indian lifestyle are not terminologically distinguished from the Indian group with whom they associate themselves.
3. While the seeds of this discontent have long existed, the recent foment in inter-group social strife coincides with the foundation of the *Consejo Aguaruna-Huambisa* and with the sequential visitations of a series of self-styled "advisers" from the 7th University of Paris.

REFERENCES

Berlin, Brent and Elois Ann Berlin. 1983, Adaptation and ethnozoological classification: Theoretical implications of animal resources and diet of the Aguaruna and

Huambisa. pp. 301–325 in *Adaptive Responses of Native Amazonians*, ed. by Raymond B. Hames and William Vickers. San Francisco and New York: Academic Press

Briggs, George and Doris Calloway. 1979. *Bogert's Nutrition and Physical Fitness*. 10th ed. Philadelphia: W. B. Saunders

Boserup, Ester. 1965. *The Conditions of Agricultural Growth: The Economics of Agrarian Change under Population Pressure*. Chicago: Aldine

Cassel, John. 1957. Social and cultural implications of food and food habits. *American Journal of Public Health* 47:732

Chumap L., Aurelio and Manuel Garcia Renueles. 1979. *"Duik Muun...": Universo Mítico de los Aguarunas*. Tomo I. Serie Antropológica II. Lima: Centro Amazónico de Antropología y Aplicación Práctica

DeWalt, Kathleen, P. B. Kelly and Gretel H. Pelto. 1980. Nutritional correlates of economic microdifferentiation in a highland Mexican community. pp. 205–222 in *Nutritional Anthropology*, ed. by Norge W. Jerome, Randy F. Kandel and Gretel H. Pelto. Pleasantville, N.Y.: Redgrave

Fitzgerald, Thomas K. (ed.). 1977. *Nutrition and Anthropology in Action*. Assen/ Amsterdam: Van Gorcum

Gifft, Helen H., Marjorie B. Washbon and Gail G. Harrison. 1972. *Nutrition, Behavior and Change*. Englewood Cliffs, N.J.: Prentice-Hall

Greene, Lawrence S. 1977. Hyperendemic goiter, cretinism and social organization in highland Ecuador. pp. 55–91 in *Malnutrition, Behavior and Social Organization*, ed. by Lawrence S. Greene. San Francisco and New York: Academic Press

Gross, Daniel and Barbara Underwood. 1972. Technological change and caloric costs: Sisal agriculture in northeastern Brazil. *American Anthropologist* 73:725–740

Guallart, José Ma. 1964. Los Jíbaros del Alto Marañón. *América Indígena* 24:315–334

Harner, Michael. 1972. *The Jivaro*. Garden City, N.Y.: Doubleday-Natural History Press

Jerome, Norge W., Randy F. Kandel and Gretel H. Pelto (eds.). 1980. *Nutritional Anthropology*. Pleasantville, N.Y.: Redgrave Publishing Co.

Lee, Dorothy. 1957. Cultural factors in dietary choice. *American Journal of Clinical Nutrition* 5(2):166–170

Loomis, Robert S. 1976. Agricultural systems. *Scientific American* 235(3):98–105

Montgomery, Edward. 1977. Social structuring of nutrition in India, pp. 143–169 in *Malnutrition, Behavior and Social Organization*, ed. by Lawrence S. Greene. San Francisco and New York: Academic Press

Sauer, Carl O. 1954. *Agricultural Origins and Dispersals*. New York: American Geographical Society

Seagraves, B. Abbott. 1977. The Malthusian proposition and nutritional stress, pp. 173–210 in *Malnutrition, Behavior and Social Organization*, ed. by Lawrence S. Greene. San Francisco and New York: Academic Press

Uriarte, L. 1976. Poblaciones nativas de la Amazonia Peruana. *Amazonia Peruana* 1(1):9–58

ADDITIONAL READING

Basso, Ellen. 1977. The Kalapalo dietary system. pp. 98–105 in *Carib-Speaking Indians: Culture, Society and Language*, ed. by Ellen Basso. Anthropological Papers No. 28. Tucson: University of Arizona

Beckerman, Stephen J. 1979. The abundance of protein in Amazonia: A reply to Gross. *American Anthropologist* 81(3):533–560

Berlin, Brent and Elois Ann Berlin. 1978. Etnobiologia, subsistencia y nutrición en una sociedad de la selva tropical: Los Aguaruna. pp. 13–48 in *Salud y nutrición en sociedades nativas*, ed. por A. Chirif. Lima: Centro de Investigatión y Promoción Amazónica

Berlin, Elois Ann and Edward K. Markell. 1977. An assessment of the nutritional and health status of an Aguaruna Jivaro community, Amazonas, Peru. *Ecology of Food and Nutrition* 6:69–81

Carneiro, Robert L. 1960. Slash and burn agriculture: A closer look at its implications for settlement patterns. pp. 229–234 in *Man and Cultures*. Selected Papers of the Fifth International Congress of Anthropological and Ethnological Sciences, ed. by Anthony F. C. Wallace. Philadelphia: University of Pennsylvania

_____ Slash and burn cultivation among the Kuikuru and its implications for cultural development in the Amazon Basin, pp. 47–67 in *The Evolution of Horticultural Systems in Native South America: Causes and Consequences, A Symposium*, ed. by Johannes Wilbert. Supplemental Publication No. 2, Sociedad de Ciencias Naturales La Salle. Caracas: Editorial Sucre

Chagnon, Napoleon A. and Raymond B. Hames. 1979. Protein deficiency and tribal warfare in Amazonia: New data. *Science* 203:910–913

Denevan, William M. 1970. The aboriginal population of western Amazonia in relation to habitat and subsistence. *Revista Geographica* 72:61–86

_____ 1974. Campa subsistence in the Gran Pajonal, eastern Peru. pp. 92–110 in *Native South Americans: Ethnology of the Least Known Continent*, ed. by Patricia J. Lyon. Boston: Little, Brown

Dricot, J. M. and C. Dricot-D'Ans. 1977. Influence des transformations socio-economiques sur l'etat de nutrition des Indiens Machiguenga (Amazonie Peruvienne): Aspects methodologiques. *Biometrie Humaine* 12:77–89

Dricot-D'Ans, C. and J. M. Dricot. 1978. Influence de l'acculturation sur la situation nutritionelle en Amazonie Peruvienne. *Annales de la Societé Belge de Medecine Tropicale* 58(1):39–48

Gross, Daniel R. 1975. Protein capture and cultural development in the Amazon basin. *American Anthropologist* 77:526–549

Hames, Raymond B. (ed.). 1980. *Studies in Hunting and Fishing in the Neo Tropics.* Working Papers on South American Indians No. 2. Bennington, Vt: Bennington College

Hames, Raymond B. and William Vickers (eds.). 1983. *Adaptive Responses of Native Amazonians*. San Francisco and New York: Academic Press

Holmes, Rebecca. 1981. *Nutritional Status in Four Tropical Rainforest Villages in the Venezuelan Amazon Territory: A Study in Adaptation and Acculturation*. Masters Thesis. Instituto Venezolano de Investigaciones Cientificas (IVIC), Caracas

Johnson, Allen. 1977. The energy costs of technology and the changing environment: A Machiguenga case. pp. 155–167 in *Material Culture*, Proceedings of the American Ethnological Society, ed. by Heather Letchman and Robert S. Merrill. St. Paul, Minn.: West Publishing Co.

Moran, Emilio F. 1974. The adaptive system of the Amazonian caboclo, pp. 136–159 in *Man in the Amazon*, ed. by Charles Wagley. Gainesville, Fl.: University of Florida Press

_____ *Developing the Amazon: Homesteaders along the Transamazon Highway*. Bloomington, IN: University of Indiana

Roosevelt, Anna C. 1980. *Parmana: Prehistoric Maize and Manioc Subsistence Along the Amazon and Orinoco*. San Francisco and New York: Academic Press

Ross, Eric. 1976. *The Achuara Jivaro: Cultural Adaptation in the Upper Amazon*. Ph.D. dissertation, Columbia University. Ann Arbor, Mich.: University Microfilms

_____ 1978. Food taboos, diet and hunting strategy: The adaptation to animals in Amazon cultural ecology. *Current Anthropology* 19(1):1–36

Siverts, Henning. 1972. *Tribal Survival in the Alto Marañón: The Aguaruna Case*. Copenhagen: International Work Group for Indigenous Affairs

Smith, Nigel J. H. 1974. Destructive exploitation of the South American river turtle. *Yearbook of the Association of Pacific Coast Geographers* 36:85–102

Vickers, William. 1976. *Cultural Adaptation to Amazonian Habitats: The Siona-Secoya of Eastern Ecuador.* Ph.D. dissertation, University of Florida. Ann Arbor, Mich.: University Microfilms

Winton, Marianne. 1970. Nutritional adaptation of some Columbian Indians. *American Journal of Physical Anthropology* 32:293–298

CHAPTER 4

STAPLES AND CALORIES IN SOUTHEAST ASIA: THE BULK OF CONSUMPTION[1]

CHRISTINE S. WILSON

Department of Epidemiology and International Health
University of California San Francisco, California 94143

ABSTRACT

For a food energy source, most low income peoples in developing regions depend on a starch staple, locally produced or available. In the Southeast Asian tropics, this food is subsistence wet rice, eaten in quantities of 1200 or more Calories per person per day by an average adult male. This amount of the staple contributes approximately 30 g of protein per day to the eater. In a Malay fishing village, about one fifth of the population obtained a portion of its rice by subsistence or barter, buying the rest. Most of the diet animal protein (30 g per person per day) came from subsistence-cum-commercial fishing activities. Manioc, yams, taro, yambean and sago were secondary carbohydrate sources taken as snacks or for breakfast. Refined carbohydrates—wheat flour and its products, sugar and sweets, particularly refined white sugar, intake of which averaged 50 g per person per day—were relatively new dietary additions, contributing to obesity and caries, and perhaps also to incidence of diabetes and hypertension.

INTRODUCTION

With few exceptions, such as the Masai and precontact Eskimo, most peoples traditionally depended for the bulk, the largest portion of their diet, upon a food energy source. This usually was a starch staple, defined nutritionally as preponderantly carbohydrate. This food may have been a grain cereal, a root crop or a tuber, grown locally or locally available. Such caloric dependence characterized most rural populations in western countries until the 20th century, when mechanized agriculture and processed, packaged foods interposed machines and commercial enterprise between western peoples and traditional ways of food-getting and eating. The memory of the physical-religious meaning of our "staff of life" has faded; the replacement manufactured substitutes lack emotional and cultural content.

For low income peoples in the developing world, continued reliance on a staple is general. Ritual activities and beliefs surround perpetuation of the crop which is central to the diet. These foods have been called cultural superfoods because of the mystical, reverential respect in which they are held by customary consumers, and for their dominant place in rites and beliefs (Jelliffe 1967).

THE ASIAN EXAMPLE

The following observations are based on research done in a coastal fishing community of 114 households (some 600 people) on the East Coast of the Malay Peninsula, in the State of Trengganu, one hundred miles south of the border of Malaysia with Thailand. The study took place in two nine-month intervals two years apart. Its intent was to learn the effects on health and nutritional status of cultural practices dealing with food and eating and to develop means of studying diet within a cultural setting that would disturb the eating milieu as little as possible (Wilson 1970:ii). The principal employment of over two thirds of the adult village males was obtaining, processing or selling sea fish commercially. These people are part of a market economy, yet obtain fish and some of their other food through their own labor. Their economy is thus subsistence-cum-commercial, a situation not untypical in developing regions. Their food production and food use were studied by methods that combined techniques of anthropological and dietary research, chiefly participant observation and interviews of housewives in the Malay language with a minimum of local assistants. Modified nutrition survey methods included this-day food recalls, observations of meal preparations and family meals, and weighing of replicate foods (Wilson 1970:172, 1978:144).

Throughout the Asian tropics the staple food is most commonly subsistence wet rice, eaten daily in amounts that provide 1600 or more Calories to an average adult male, who may eat the equivalent of a pound of raw rice each day. Other carbohydrate staples also contribute to Asian diets, chiefly as snacks taken between meals or for breakfast. These are widely grown in Asia: tapioca (*Manihot utilissima*, elsewhere called manioc or cassava; Burkill 1966:1434), various species of yam (*Dioscorea*, Burkill 1966:824 ff.), taro root (*Colocasia esculentum*, Burkill 1966:647), yambean (*Pachyrrhizus erosus*, Burkill 1966:1647) and sago (*Metroxylon sagus*, Burkill 1966:1484). Manufactured wheat products, such as imported or locally made bread and sweet and plain biscuits (cookies and crackers) are additional carbohydrate snack foods throughout Asia.

In Southeast Asia, homemade fried cakes and pancakes from imported wheat flour are further sources of carbohydrate calories. They sometimes take the place of more usual fried leftover rice or rice cakes as the first meal of the day. Other common snack or breakfast foods adopted by non-Chinese Southeast Asians are dishes based on noodles (*mi*, pronouned mee, from the Chinese *misoa* [Winstedt 1965:237]) manufactured locally from wheat flour. Noodles from rice flour, termed *beehoon* in Malay, are also eaten. Westerners call them glass noodles.

The general eating pattern in this region is a snack breakfast followed by two rice meals, each consisting of large servings of rice with much smaller accompaniments of dishes made from some animal protein source or vegetables (called *luuk* in Malay), and still smaller quantities of relishes (*sambal*, in Malay). Since there is also considerable between-meal nibbling, carbohydrate foods indeed bulk large in the diets of Southeast Asians.

MALAY USES OF CARBOHYDRATES

In much of Southeast Asia, fish are more common fare than meat or poultry. In the East Coast Malay fishing village studied, fish and other sea foods are well-liked items of diet. Infrequent feasts substitute chicken or prestigious meat for fish. With this exception, a meal lacking fish is not well received, and fish are the principal source of animal protein for these Malays. In the amounts eaten they provide approximately the same quantity of protein to the diet as does rice, 30 g (1.05 oz.) or more per person per day. It is worth nothing that, although the amount of protein in rice is low, averaging some 7 g per 100 g of raw rice, the quality of rice protein is quite good, better than that of either wheat or maize (FAO 1954:13,65). This is because the

TABLE I

Calories available from purchased rice and sugar in selected Malay households[a]

Household[b]	Number Persons Eating	From Rice Cal	(mJ)	From Sugar Cal	(mJ)	Total Cal	(mJ)
Othman	7	897	(3.75)	50	(.21)	947	(3.96)
Idris	6	524	(2.19)	133	(.56)	657	(2.75)
Razali	5	1256	(5.25)	138	(.58)	1394	(5.83)
Mohamed	5	335	(1.40)	96	(.40)	431	(1.80)
Yusof	6	1047	(4.38)	200	(.84)	1247	(5.22)
Abdul Hamid	5	628	(2.63)	96	(.40)	724	(3.03)
Yahya	12	436	(1.82)	80	(.34)	516	(2.16)
Daud	7	2692	(11.28)	228	(.95)	2921	(12.22)
Nasser	6	1047	(4.38)	200	(.84)	1247	(5.22)
Muda	6	2094	(8.76)	400	(1.67)	2494	(10.43)
Hussein	5	1256	(5.25)	240	(1.00)	1496	(6.25)
Omar	3	628	(2.63)	227	(.95)	855	(3.58)
Ibrahim	10	1047	(4.38)	480	(2.01)	1527	(6.38)
Ali	6	1047	(4.38)	200	(.84)	1247	(5.22)
Salleh	4	1047	(4.38)	200	(.84)	1247	(5.22)
Ismail	3	1047	(4.38)	400	(1.67)	1447	(6.05)

[a] Wilson (1970:167).
[b] Pseudonyms for heads of households.

amino acids present in rice are in about the proportions that would be needed by human beings for protein synthesis, with one exception. As is the case with the other two cereals, the essential amino acid lysine is limiting in rice protein. This means that, in comparison with the protein needs of the human body for eight amino acids it cannot synthesize, only lysine is not present in rice in sufficient proportion to make rice protein comparable in amino acid quality to some animal protein such as egg (FAO 1954:13).

Fish is a good source of the lysine in which rice is inadequate (Chong and Soh 1966). In Asia, rice and fish are usually taken together at the same meal. Amino acids must be ingested at about the same time in order to be available for protein synthesis within the body. Therefore, most Southeast Asian subsistence rice eaters who have access to fresh fish have, despite a high starch diet, a relatively good amino acid intake pattern and, therefore, protein intake, and their nutritional status with respect to this nutrient is usually satisfactory (Wilson 1970:439, 497).

Increasingly, another major carbohydrate calorie source for village dwellers everywhere is white sugar. Sugar cane is grown in Malaysia. The cane is chewed as a snack food or squeezed in hand-operated presses to make a drink. Since the middle of the 20th century refined white sugar has been imported into the country. This has largely replaced the brown sugar made locally from the inflorescence of the coconut palm, *gula Melaka* (Malacca sugar), which was consumed only in small amounts (Firth 1966:99). White sugar (*gula*) is now also manufactured in Malaysia.

The amounts of refined white sugar purchased for household use in the study village would provide an average intake of 50 g (1.76 oz.) per person per day for every man, woman, child and infant in the 16 households for which calculations were made, providing an added 200 Calories (0.84 MJ) to each person, each day (Table I). This amount represents 40 pounds (18 kg) per person per year. These data may be compared with figures given for the United States by the Department of Agriculture, which indicated an average American consumed approximately 129 pounds (59 kg) of sugar in 1978 (American Public Health Assn. 1980). However, 75% of this total (about 97 pounds, or 44 kg) was nondiscretionary, being added to food and drink during processing. One may conclude, therefore, that 32 pounds (14.5 kg) of sugar as such was eaten by each average American. In contrast, the figures given for Malay village sugar consumption refer to what is termed visible or discretionary sugar. It is generally introduced only into drinks—tea or coffee. Malay women use little sugar in cooking. Most of this sugar is consumed by adults, since stimulant drinks are infrequently offered to children.

Comparable national information on nondiscretionary sugar consumption for Malaysia is not available. Statistical data from the Ministry of Agriculture and Cooperatives (1968:44, 84) do indicate that a net of 93,397 tons (84,906 metric tons) of "pastries, cakes and bakers' wares," 1,334,759 pounds (606,709 kg) of ice cream and ice cream powders, and 92,047 gallons (348,426 L) of flavored, non-alcoholic beverages were imported into the western portion of the country (excluding Sarawak and Sabah) in 1966, two years before the reported study commenced. Divided by the population of West Malaysia at this period, 8,415,488 (Anon. 1969:17), it may be calculated that each resident of the Peninsula had a potential intake of 25 pounds (11.4 kg) a year of pastries and sweets, but only 72 g (2.5 oz) of ice cream and 0.175 cups of presumably sweetened beverages. How many additional quantities of these products were made in the country for domestic consumption the available statistics do not indicate, but food processing is one of the industries being developed in Malaysia. Consumption within the village of commercially produced sweet foods was determined for only a few individuals. This appeared to corroborate the aforementioned annual estimate of 25 pounds (11.4 kg) per individual for the country as a whole, and would represent about 32 g, (1 oz) or 140 Calories, for each person per day.

HOUSEHOLD ECONOMICS AND CALORIES

In the course of a household census initiated early in the study village, queries were made regarding land ownership and use, and regularly recurring purchases of foodstuffs, staples, fruits and vegetables, and other nonfood necessities. These interviews revealed that a majority of families either own some rice land or engage in other agricultural activities, such as raising coconuts, fruit trees or a few vegetables. Depending upon the extent of their economic involvement in fishing-related activities (the single largest source of income for the community), and the amount of suitable land owned, one sixth of the 114 households in this village were able to obtain a portion of their annual consumption of rice by raising it themselves, or by bartering goods or services for it. The amount so obtained varied from about 40 to 80 gallons (151 to 302 L) unhusked rice (*padi*), which represents 17.5 to 35 gallons (166 to 132 L) of husked rice (*beras*) (Wilson, unpublished observations, 1984). The rest of their needs was purchased from local shops or nearby markets. Some of the rice was imported from Thailand, only 100 miles (160 km) to the north. The white, highly milled Siamese rice (*beras Siam*) is greatly prized. Home-grown rice is pounded in a mortar or husked by small gasoline-driven mills to a lower degree of milling.

Household purchase figures were used to determine amounts of rice, sugar and cooking oil consumed daily. Most purchases were reported as monthly or fortnightly acquisitions. Malay housewives buy larger quantities at lower unit prices when income permits. Calculation of calories available from these staple foods was made on a *per caput* basis for all family members except nursing infants less than one year of age. Confirmation of amounts of rice consumed was obtained from certain households by querying the usual number of measures of raw rice cooked for one or two meals—a day's quota for the family. The measure, a round cigarette tin, was readily obtainable to replicate and weigh amounts stated.

TABLE II

Principal caloric sources for a Malay village

	Amount/Household/ Year[a]	Energy Intakes Person/Day Cal (mJ)	% Malaysian RDA[e]	% FAO/ WHO RDA
Rice (husked, uncooked)	182–234 gantang[b]	1200–1500 (5.02–6.28) (1350 av.) (5.65)	Males: 63.4 Females: 79.6	68.4 90.1
Granulated Sugar	190 kati[c]	200 (.84)		
Cooking Oil	45 chupak[d]	180 (.75)		
Total Cal (mJ) from rice, sugar, oil		1730 (7.24)		

[a] Average household size is 5.2 individuals.
[b] A gantang = one gallon, 3628.8 g of rice.
[c] One kati = 1-1/3 pounds (0.6 kg)
[d] A chupak = one quart
[e] RDA = Recommended Dietary Allowance for adults

In 7 of 16 households, daily rice consumption per head was so calculated at 300 g, (10.6 oz), providing 5.0 MJ (1047 Cals) (University of Malaya, unpublished observations).[2] In the other 9 households studied it ranged from 900 to 1200 Cals (4.3–5.7 MJ) (Table I). Preadolescent children made up nearly half the almost 100 individuals in this subsample and would not be expected to consume as much rice as an average adult. It thus becomes apparent that these adult villagers rely heavily on their carbohydrate staple for energy. From rice eaten at home, adult males in this village could obtain up to one half the FAO/WHO (1973:31) recommended energy intake of 2530 Cals (12.1 MJ) for a moderately active man weighing 55 kg (120 lb), the approximate body size of these rather small and slight Malay fishermen. Calories derived only from rice, refined sugar and cooking oil, consumed in home meals by a hypothetical adult in this community, may be estimated to total more than 1500 Cal (7.2 MJ) (Table II).

Subsequent observation of mealtimes in some households, and comments from villagers, strongly suggest that daily intakes of rice are stable and predictable. Indeed, in the second nine-month phase of study in this village, inquiries made of families newly moved into the community regarding food purchases, and an economic study of a sample of 32 families randomly selected from the village population, both indicated that rice consumption closely resembled that determined two years earlier in quantity and caloric contribution (Wilson, unpublished observations).

Amounts of rice and other carbohydrate foods taken as snacks away from home were more difficult to quantify. For men, who usually eat breakfast and other snacks at one of the local coffee shops, estimates of calories from these sources were made from shopkeepers' records of sales. For example, a nearby shop served about 50 cups of coffee or tea per day. The amount of sugar and sweetened condensed milk contained in these drinks provides each consumer with 220 Calories. The number of cakes this shop sold, divided by the same number of customers, contributed another 290 Calories per man. Thus most adult males obtained a fifth to a quarter of their day's calories from these shops, carbohydrate foods again being the overwhelming contributor.

RICE GROWING

Obtaining rice by one's own effort was a comparatively new activity in this village, and plots were for the most part small, ranging in size from about one quarter to one half acre (0.1 to 0.2 ha). Despite the predominance of sandy soil due to closeness to the ocean, rice grew satisfactorily. A few empty bunded fields testified to past efforts at rice raising. Due to lack of irrigation facilities, only one rice crop was grown annually, when the winter monsoon provided the natural flooding needed by the crop. Upland or dry rice was also raised, but by only two households, headed by elderly people, each of which had only a quarter acre (0.1 ha) or less, suitable for this purpose.

More village families (about 15%) had rice land in an inland part of the state, some 20 miles (32 km) distant, in the chief rice-growing area of Trengganu, near the Trengganu River, which empties into the South China Sea 12 miles (19 km) north of the study village. This area could be reached by bus; only two families in this community had automobiles, although many owned bicycles. The Malaysian Government made the land available for purchase, in lots approximately one acre (0.4 ha) in area, in the early 1960s. Payment was made in installments by turning over one third to one half the crop to the government each year or by annual cash payments.

Figure 1. Transplanting rice seedlings from nursery plot (foreground) to flooded *padi* fields behind, Trengganu, Malaysia.

Double-cropping is practiced in this part of Trengganu, but those from the study village who owned land in this part of the state commented that the second crop, dependent on irrigation water in the dry season, required very hard work for a smaller yield. Few of them planted a second crop. Raising rice or any other crop represents an investment in time not only for seeding, transplanting (Figure 1) and weeding, but also in harvesting, winnowing, storage and milling and this conflicts with the traditional preoccupation with the sea. Nearly 70% of males over the age of 15 years were actively and almost fully engaged in fishing or enterprises closely related to it. Despite this, by the time of the second nine-month study, almost a fifth of village households were obtaining part of their rice supply from agricultural efforts.

The majority of the agricultural work, except preparation of the soil for planting, is done by women. No time schedules were collected for these agricultural activities. Those who live within walking distance of their fields sow, transplant and weed in time taken from other household tasks. In the chief rice-growing area of Trengganu, men sometimes help to pull up the seedlings and carry them in bundles from the nursery plots to the plowed fields for transplanting. For varieties which may be harvested by sickle, men as well as women cut the sheaves and carry them to the threshing box where they are beaten against its sides to dislodge the grains, which fall into the container below. This activity is called *pukul padi*, *pukul* meaning "to hit" (*padi* is rice, growing or harvested, before it is milled). Other varieties of rice are still harvested by women, as they have been for centuries in all of Malaysia, a single stalk at a time, using the *tuai*, a small reaping knife held in the palm of the hand to avoid frightening the spirit of the rice (*semangat*), which Malays believe inhabits every plant (Skeat 1900:227). Rice harvest by sickle or scythe, encouraged by the authorities, appears to be a modern introduction in Malaysia (Skeat 1900:227). This practice did not begin in Trengganu until 1970 (Wilson, unpublished observations). A number of different varieties of rice may be planted in the same plot. Since all do not mature at the same time, use of the *tuai* for harvesting is more functional than that of the sickle in this case.

A woman with a quarter acre (0.1 ha) plot in the village spent nearly two weeks harvesting alone with the *tuai*, although her husband, a prosperous fisherman, had earlier indicated he intended to rent an *enjin* to accomplish the task. (She also continued to attend to those household chores she felt her three preteenage daughters were not equipped to do.) Others who collaborated with kin and friends (the act is called *tolong-menolong*, to give and receive help) to harvest larger lots in more distant locales would be gone from their home village a week or ten days to complete the work, staying with a relative for the duration of the harvest. Husbands and school-age children remained at home, with girls above 10 or 12 doing the cooking.

Reaping began early in the morning and usually continued until nearly sundown. Most workers took a midday break of an hour or so in a shelter or under a tree near the fields to eat picnic-style from *tiffin* containers of food brought from a relative's kitchen.

In the larger, inland rice-growing area, the harvested rice could be milled in a gasoline-driven machine for a small fee. A resident of a local village ran this mill. Hand husking required 15 minutes of labor to pound the grains in a large wooden mortar (*lesong*) with a long, heavy wooden pestle for each quart (0.9 L) of raw rice obtained. Because more of the germ remains, hand-hulled rice has greater nutrient value than rice that is machine-milled. Hand-husked rice is also more flavorful. Some housewives were willing to expend the extra time required to obtain tastier, home-pounded rice.

Reported yields were variable. They may reflect the variety of rice planted, or the time that could be spent tending distant fields. Neither weed killer nor pesticides were reported to be in use in this part of Malaysia during the research period. From two inland acres (0.8 ha) one family got 200 *gantang* (gallons) (720 L) of *padi* (raw unhusked rice) in the first nine-month period, 80 (288 L) in the second. Another household realized 200 *gantang* of *padi* from 2.5 acres (1 ha). From 0.25 acre (0.1 ha) in the village a third family reaped 40 *gantang* (144 L). The yield of rice (*beras*) after milling is one half or less the weight of the unhusked *padi*. At an estimated daily raw rice intake of 300 g (11 oz) per person per day, an annual individual intake would be 30 *gantang* (108 L) of raw, husked *beras*. If the daily intake were 400 g (14 oz) the year's total intake would be 40 *gantang* (144 L) per person. Yields higher than these were not reported for this village. Thus those fishermen with larger yields obtained from household agricultural efforts enough rice for up to two members of their families for one year, or, for an average-size family (5.2 individuals), a third or less of their annual rice needs (Table II).

Other means of obtaining rice were in exchange for helping to harvest or by bartering rice sacks woven from pandanus (screw pine, *Pandanus tectorius*, Burkill 1966:1678). These were also chiefly female activities. One particularly enterprising fisherman's wife used both means. She reported being paid 50 *gantang padi* (180 L) for 11 rice sacks she had woven for this purpose.

A special variety of rice is bought for making into cakes and other festival foods. It is glutinous or "sticky" rice, called *pulut* in Malay. Although it is about twice as expensive as ordinary *beras*, a few families with higher incomes bought it two or three times a month, steaming or boiling it to serve for breakfast, sprinkled with grated coconut, salt and sugar. Most households use it for feasts or foods for other holidays, which occur less frequently than once a month. It is the usual base for rice wine, not consumed in this Muslim community. Since consumption of *pulut* rice in any form is intermittent, it has not entered into the calculations presented here. When eaten, its calories substitute for, rather than supplement, those from ordinary day-to-day foods (Wilson, unpublished observations).

OTHER CARBOHYDRATE FOODS

Root crops, tapioca, yams, or taro, grown in the dooryards or beside the nearby well make less frequent contributions to the diet. There is a tendency to eat them as an occasional vegetable side dish with rice, or between meals alone as a snack. However, tapioca (*ubi kayu* which means "wood root"), the commonest and most prevalent root vegetable, never substituted for rice in any main meal observed or reported during the study, despite reports by nutritionists and others that it is so used among Malay and Indonesian rural dwellers (van Veen 1971:30). Instead it was served as breakfast, or offered to visitors as a snack accompanying coffee. Those households which grew it planted it no oftener than once a year, then waited approximately six months for the roots to develop sufficiently to be eaten. Malays either boil or roast them in an open fire in their skins. Both processes inactivate the prussic acid they contain. Once cooked, tapioca contains chiefly carbohydrate, with at best traces of other nutrients. A common use of *ubi kayu* was as a fermented cake called *tapai*, made from sliced boiled roots and local yeast (Wilson, in press). Some East Coast housewives made them in quantity for sale to other housewives in plastic sacks, ten cents Malay (with equivalent U. S. buying power) for a dozen, at twice-weekly markets. (The nearest market to the study village was a mile distant.) This delicacy is made in the same manner with steamed glutinous rice for holidays, but apparently is never sold. These rice cakes were gifts, instead, to kith and kin.

Other starch staples which, unlike tapioca, contain minerals and other nutrients were less common in the diets of these villagers. Yams (*Dioscorea alata*, called *ubi*, meaning "root," in Malay), though economically important native tubers (Burkill 1966:826 ff.), were much less frequent diet items for these villagers. They require about 10 months after planting to become edible. Taro (*Colocasia esculentum*) and *Xanthosoma vioaceum*, known as *yautia* in the New World (Holttum 1954:54, Burkill 1966:649, 2312), are both called *ubi keladi* by these East Coast Malays. They contain needlelike crystals of calcium oxalate, recognized by Malays as needing removal by peeling and cooking well before consumption, and perhaps for this reason they are not often eaten. Sweet potatoes are occasionally taken as a snack food (as is maize). They are made into a pudding or similar dish. Sage (*Metroxylon sagus* or *M. Rumphii*, Burkill 1966:1484) is used primarily as flour that is mixed with cooked fish to make *krupuk*, "chips" that are sundried before frying. The tuberous root of the yambean (*Pachyrrhizus erosus*), a leguminous climber called *bengkuang* in Malay, is sometimes sliced and eaten raw as a relish accompanying a meal, in the same manner in which cucumber is served in this region.

FISH DISBURSAL

Part of the fish catch is commonly held back from sale for routine sharing as partial (kind) payment among boat crews (Figure 2). The fisherman may sell his share, or take it home to his family to eat. Nonfishing families usually obtain their fish as gifts or by purchase from fishermen kin. When commercial buyers (represented by middlemen) offer high prices for the fish, the captain of the boat may decide to sell the entire catch. He will then join his crew in competing with the middlemen to buy food fish from other boats.

When catches are bountiful and not all fish are sold (for example, if the middleman's offer is too low), housewives may dry excess fish for consumption during the monsoon season when seas are rough and smaller boats are beached for safety.

There were a number of fish-drying businesses in the community, commercial enterprises that shipped their wares via lorry to the cities,

Figure 2. Distribution of part of the fish catch among the crew (squatting men) of a Malay fishing boat.

and even to countries outside Malaysia. They also sold smaller quantities of dried fish to local housewives during the stormy season.

DISCUSSION AND CONCLUSIONS

Evidence of a generally good protein status in this particular village population was shown by anthropometric measurements, made by a pediatrician colleague, of midarm circumference (Wolenski, quoted by Jelliffe 1966:228) and triceps skinfolds on 33 preschool children. Use of a formula developed by Jelliffe (1966:77) that determines midarm muscle mass from the arm circumference and skinfold measures taken at the same point showed significantly greater muscle mass in these seacoast children than in Malay Army dependents or children living 25 miles (40 km) inland where fish were eaten infrequently. The results in the study village were a direct reflection of good protein intake (Wilson 1970:495). This finding was confirmed by laboratory tests of blood specimens. Blood samples taken from 17 children and adults in the community had entirely satisfactory levels of total protein (measured by microkjeldahl procedures, Wootton, 1964), ranging from 6.7 to 8.8%, with a median of 7.4 and a mean of 7.5 (Wilson 1970:440–441).

Starchy staples and fish have been the diet bases for this population for centuries, with little apparent detriment to health. Refined carbohydrates represented by sugar, biscuits and sweets are newer introductions, commercially refined sugar becoming available in quantity only after World War II. Increased consumption of these last items may have contributed to some obesity observed in this community, and is probably responsible for widespread, readily evident tooth decay. One may also be concerned about the possible role of these manufactured foodstuffs in diabetes and hypertension, both conditions the villagers themselves recognize and know as reasons for health anxiety. Traditional staples formerly contributed about two thirds of total diet calories. The newer carbohydrate foods now represent a fifth or more of energy intakes, but provide less of other needed nutrients than the older foods. Consumption of less highly milled rice grown by personal effort may slightly offset this disadvantage. Certainly, in a community so geared and oriented to the sea, occurrence of these relatively new agricultural activities should encourage planners who hope to introduce diversification into village populations.

NOTES

1. The research on which this paper is based was supported by the following grants from the U. S. National Institutes of Health: GM 35001-3 and AM 19152-01 to the author; and AI 10051 to the Department of Epidemiology and International Health, University of California, San Francisco, through the U. C. International Center for Medical Research (UC-ICMR) at the Institute for Medical Research, Kuala Lumpur, Malaysia.
2. University of Malaya (1968). Food Composition Tables Compiled for Use in West Malaysia. Department of Social and Preventive Medicine, Faculty of Medicine.

REFERENCES

American Public Health Association. 1980. Much of Sugar Pre-Packaged into Foods. *The Nation's Health*, January. Washington: APHA
Anonymous. 1969. *Malaysia Yearbook 1968/9.* Kuala Lumpur: The Straits Times Press
Burkill, Isaac H. 1966. *A Dictionary of the Economic Products of the Malay Peninsula.* 2 vols. Kuala Lumpur: Ministry of Agriculture and Cooperatives
Chong, Y. H. and C. C. Soh. 1966. The protein quality of ikan bilis (*Stolophorus* spp.). *Medical Journal of Malaya* 22:330-333

Food and Agriculture Organization. 1954. *Rice and Rice Diets. A Survey Prepared by the Nutrition Division*. FAO Nutritional Studies No. 1 Rome: FAO

FAO/WHO. Joint *ad hoc* Expert Committee. 1973. *Energy and Protein Requirements*. WHO Technical Report Series No. 522, FAO Nutrition Meetings Report Series 52. Geneva: FAO/WHO

Firth, Rosemary. 1966. *Housekeeping among Malay Peasants*. Second edition. London: Athlone Press

Holttum, Richard E. 1954. *Plant Life in Malaya*. London: Longmans, Green

Jelliffe, Derrick B. 1966. *The Assessment of the Nutritional Status of the Community* (with special reference to field surveys in developing regions of the world). World Health Organization Monograph Series No. 53 Geneva: WHO

―――― 1967. Parallel food classification in developing and industrialized countries. *American Journal of Clinical Nutrition* 20:279–281

Ministry of Agriculture and Cooperatives (Kementarian Pertanian dan Sharikat Kerjasama). 1968. *Statistical Digest (Rumusan Perangkaan)*. Kuala Lumpur: Planning and Research Division (Bahagian Peranchang dan Penyelidekan)

Skeat, Walter William. 1900. *Malay Magic. Being an Introduction to the Folklore and Popular Religion of the Malay Peninsula*. London: Macmillan

Veen, Marjorie S. van. 1971. Some ecological considerations of nutrition problems on Java. *Ecology of Food and Nutrition* 1:25–38

Wilson, Christine S. 1970. *Food Beliefs and Practices of Malay Fishermen: An Ethnographic Study of Diet on the East Coast of Malaya*. unpublished Ph.D. dissertation, University of California, Berkeley

―――― 1978. Developing methods for studying diet ethnographically. pp. 141–148 in *The Anthropology of Health*, ed. by Eleanor E. Bauwens. St. Louis: C. V. Mosby

―――― 1985. Nutritional and social contexts of "ethnic" foods: Malay examples. in *Shared Wealth and Symbol: Food, Culture and Society in Oceania and Southeast Asia*, ed. by Lenore Manderson. Cambridge: Cambridge University

Winstedt, Richard O. 1965. *An Unabridged Malay-English Dictionary*. Edition 6. Kuala Lumpur: Marican & Sons

Wootton, I. D. P. (ed.). 1964. *Micro-Analysis in Medical Biochemistry*. Edition 4. London: Churchill

ADDITIONAL READING

Burling, Robbins. 1965. *Hill Farms and Padi Fields: Life in Mainland Southeast Asia*. Englewood Cliffs, N.J.: Prentice-Hall

Coedes, George. 1968. *The Indianized States of Southeast Asia*. English translation by Susan Brown Cowing. Honolulu: East-West Center Press

Cowan, Charles D. 1961. *Nineteenth-Century Malaya. The Origins of British Political Control*. London Oriental Series, Volume 11. London: Oxford University Press

Dobby, E. H. G. 1955, 1957. Padi Landscapes of Malaya. *Malayan Journal of Tropical Geography* 6, 10

Firth, Raymod, 1966. *Malay Fishermen. Their Peasant Economy*. 2nd Edition. London: Routledge and Kegan Paul

Fraser, Thomas M., Jr. 1960. *Rusembilan: A Malay Fishing Village in Southern Thailand*. Ithaca, New York: Cornell University Press

Hall, Daniel G. E. 1955. *A History of Southeast Asia*. London: Macmillan

Ooi, Jin-Bee. 1963. *Land, People and Economy in Malaya*. London: Longmans, Green

Swift, M. G. 1965. *Malay Peasant Society in Jelebu*. Monographs on Social Anthropology No. 29. London: Athlone Press

Whyte, Robert O. 1974. *Rural Nutrition in Monsoon Asia*. New York: Oxford University Press

Wilson, Christine S. 1973. Food taboos of childbirth: The Malay example. *Ecology of Food and Nutrition* 2:267–274.

_____ 1981. Food in a Medical System: Prescriptions and Proscriptions in Health and Illness among Malays. pp. 391–400 in *Food in Perspective*. ed. by Alexander Fenton and Trefor M. Owen. Edinburgh: John Donald

Wilson, Peter J. 1967. *A Malay Village and Malaysia. Social Values and Rural Development*. New Haven: HRAF Press

Winstedt, Richard O. 1961. *The Malay Magician, Being Shaman, Saiva and Sufi*. Revised edition. London: Routledge and Kegan Paul

Wolff, Robert J. 1965. Meanings of food. *Tropical and Geographic Medicine* 17:45–51

CHAPTER 5

A MOSAIC OF TWO FOOD SYSTEMS ON PENANG ISLAND, MALAYSIA[1]

EUGENE N. ANDERSON. JR.
Department of Anthropology
University of California
Riverside, California 92521

ABSTRACT

Two sharply different agricultural systems, with different ecological adjustments, coexist in a complex mosaic on Penang Island, Malaysia. Separation is maintained by ethnicity. Malays grow rice and tree crops, with some manioc and vegetables, re-creating a tropical forest environment that is stable with relatively little additional input. Chinese engage in intensive vegetable and livestock rearing, with fewer tree crops and virtually no rice; they produce for the market and depend on expensive inputs such as fertilizers to maintain their system. An optimal agricultural system for the island would include elements of both these systems, but ethnic tension and rivalry prevent such a fusion.

INTRODUCTION

Cultural ecologists and agricultural economists have often told us that there is usually a good adjustment between a traditional farming scheme and its environment. Indeed, many cultural ecologists (for example Anderson 1972) have been accused of a Panglossian belief that every traditional culture reached a best possible fit with its environment, or at least that all human behavior anent the ecosystem was rational in the narrow material sense and that any and all traditional or culturally established behaviors anent the ecosystem

83

were adaptive and functional. This view is identified especially with the writings of Marvin Harris (1966, 1968, 1979). Recently there has been some tendency to abandon this approach and look to process and change instead (Vayda and McCay 1975; McCay 1978), or to the importance of conflict and struggle in modifying ecological adjustments (Diener and Robkin 1978; Diener, Nonini and Robkin 1978). This paper presents a case in point. Two farming systems have coexisted for over over a hundred years on Penang Island, Malaysia, as in much of the rest of southeast Asia. They have influenced each other but have remained separate systems. Is either a "best case" fit to the tropical environment? Is one group perversely refusing to accept a better strategy? I argue that both are good but suboptimal solutions to local problems and that the differences are maintained by social and economic factors rather than by natural aspects of the ecosystem. I use the concept of a food system as being a total system of production, distribution and consumption of food. My usage implies that to understand any aspect of the food system of a culture, reference to its other components is usually necessary. This conceptually contrasts with the customary separation of food studies into those concerned with production and those with consumption.[2]

The two farming systems discussed are those of the Malays and the Chinese, respectively. The Malays practice a traditional southeast Asian agriculture based on rice and treecrops. The Chinese practice intensive vegetable and livestock raising. In rural Penang, as elsewhere in Malaysia, these two ethnic groups are numerous. There are Chinese communities and Malay communities, socially more or less segregated, but spatially almost randomly interspersed. Similar conditions exist in many other southeast Asian countries; long-established local groups practice rice-tree crop farming while immigrants are market gardeners, the two patterns existing in fine-grained mosaic.

Thus rural Penang is a mosaic of two agricultural systems. These are but the production aspects of two food systems. Chinese and Malay food preferences and diets are different and remain different as decades go by, with little acculturation or assimilation on either side. The agricultural systems that produce their food, therefore, remain separate because of demand. Economic logic would lead us to expect that the cheapest foods would drive the others from the fields, but that is not the case. This paper explores the reasons for this apparently irrational situation. First, the two systems will be described, then they will be compared, and finally a more economically rational system will be tentatively outlined.

THE MALAY FOOD SYSTEM

Traditional Malay agriculture consists of extensive rice-growing in lower, easily irrigated areas, combined with vegetable and tree horticulture on higher, better-drained sites. Various wild plants are used from both sites. Chickens are the commonest livestock, but goats, turkeys, ducks and water buffaloes occur. Water buffaloes were traditionally used as plow animals and later as meat; their replacement by small cultivating machines has reduced the protein supply in some areas. Agriculture is complemented by fishing-almost entirely marine. Malay villages are located near the shore when possible, and fish rather than livestock provide most of the animal protein. Villages are also located among the trees. The air space over the houses produces food, while the houses, set up on piles, do not interfere with the extension of tree roots beneath them. The rice land is always nearby. Ideally, a village will be situated where a stream enters the sea and will be equally close to riverine, marine, upland and lowland environments on the border between well-drained sites and paddy land.

The principal crop, wet-grown rice, can yield more calories per acre than any other grain. Malay rice-farming in the Penang area is intensive and has benefited from Green Revolution technology. High-yield seeds and commercial fertilizers and pesticides are generally used. Manioc, even higher in calorie-per-acre yields but much less high in protein and vitamins per pound or per calorie, is a back-up staple grown on higher ground around the houses. Sugar cane, also high in calorie yield but low in nutrients, is an important crop. Most sugar on Penang is bought, though it is raised and refined nearby. The home-raised cane is used for chewing and for juice (sap).

Perhaps surprising to those more familiar with temperate zones is the high productivity of the mixed tree crops. Coconut-banana intercropping can produce 6000 pounds of edible material per acre, about as much as paddy rice does in Penang, while coconut-manioc can do even better, and other coconut-intercrop systems are similarly productive (Child 1964). Coconut is the principal source of fat in the Malay diet. Mature coconut meat has some 3.2 g of protein and 28.2 g of fat per 100 g of edible meat. For Malay use it is usually grated, soaked and pressed to extract a cream-like liquid used in making curries, puddings and sweets. Coconuts, whose fibrous roots do not spread widely and thus do not compete strongly with other trees, are the main tree crop of the villages of Penang. (On the hills, where the soil is poorer, rubber dominates everywhere; it is an economic mainstay of

Malaysia but not a food crop and is not discussed here.) In addition to coconuts and bananas, tree crops commonly grown include guava (*Psidium guajava*), papaya (*Carica papaya*), mangosteen (*Garcinia mangostana*), rambutan (*Nephelium lappaceum*), durian (*Durio zibethinus*), cashew (*Anacardium occidentale*), mango (*Mangifera indica*) and occasionally other local species such as jackfruit (*Artocarpus integrifolia*) (See Table I.). These are typically grown in dooryards and also in orchards and forest clearings. *Terminalia catappa* and the rose-apple (*Eugenia jambos*) grow on beaches. The betel-nut or areca palm (*Areca catechu*) is common enough to name the island—*pinang* is the Malay name for both. Wild resources of importance include the nipa palm (*Nypa fruticans*); its fruit is eaten, although the palm is more important as a source of thatching leaves. Vegetables in addition to manioc are primarily eggplants, chiles, tomatoes and okra or ladyfingers (*Hibiscus esculentus*). Pineapples, black pepper and other minor plants occur, and most households have sweet pandanus (*Pandanus odoratissimus*) for flavoring confections and drinks.

The tree-crop combinations of the Malay food system recapitulate the complex, species-rich, multistoried structure of the lowland, tropical forest. A tree-crop combination protects the soil from the destructive effects of tropical sun and rain; reduces the chance of pest epidemics makes maximal use of light and of soil nutrients; provides shade and fuelwood; and provides habitat for domestic and wild animals. Rice paddies also conserve the land base by trapping silt and by providing a medium for nitrogen-fixing, blue-green algae by which soil fertility is maintained.

The Malay system is a conservative one in both senses of the word. It has not changed much over the centuries, and it conserves resources. Not only soil, but also labor and capital are used efficiently and sparingly. Small plots of land support families without large capital inputs or continuous heavy toil. There are probably no agricultural systems anywhere in the world that supply more food over a longer time for less factor inputs. The farms do take considerable work, but able-bodied men are apt to fish or do wage work in nonfarming hours. A well-organized Malay farm can run without continual heavy labor input, although inputs may have to be very high at rice transplanting and harvesting times. Even in the ricefields, though, the tropical climate allows staggered planting, such that small plots are continually coming into harvest rather than all the plots ripening at once and thus creating an almost impossible situation for labor supply.

TABLE I

Main Malay and Chinese Crops of Penang Island[a]

Staple crops	Scientific name	Spices	Scientific name
coconut	Cocos nucifera	betel nut (areca)	Areca catechu
manioc	Manihot esculenta (= utilissima)	chile	Capsicum annuum & Capsicum frutescens
rice	Oryza sativa	cloves	Eugenia caryophyllus
		ginger[b]	Zingiber officinale
Fruits		nutmeg & mace	Myristica fragrans
banana[b]	Musa spp.[c]	black pepper	Piper nigrum
cacao	Theobroma cacao	turmeric	Curcuma domestica
cashew	Anacardium occidentale	Other crops	
chiku	Achras zapota	amaranth greens	Amaranthus sp.[b]
citrus	Citrus spp.[c]	Chinese mustardgreen[b]	Brassica chinensis
durian	Durio zibethinus	Cucumber	Cucumis sativus
guava	Psidium guajava	eggplant	Solanum melongena
jackfruit	Artocarpus integrifolia (= heterophyllus)	gourds[b]	Lagenaria siceraria, Luffa spp., Momordica charantia
langsat	Lansium domesticum	horseradish tree	Moringa oleifera
mango	Mangifera indica	lemon grass	Cymbopogon citratus
mangosteen	Garcinia mangostana	okra	Hibiscus esculentus
papaya	Carica papaya	onion	Allium cepa
pineapple	Ananas comosus	flavoring pandanus	Pandanus sp.
rambutan	Nephelium lappaceum	peanut[b]	Arachis hypogaea
rose apple	Eugenia jambos	squashes[b]	Cucurbita moschata
(& close relatives)	Eugenia spp.	sugar cane	Saccharum officinarum
soursop	Annona muricata	sweet potato[b]	Ipomoea batatas
sweet pandanus	Pandanus odoratissimus (= tectorius)	taro[b]	Colocasia esculenta
singapore almond	Terminalia catappa	tomato	Lycopersicon esculentum
tamarind	Tamarindus indica	yard-long beans[b]	Vigna sinensis
		rubber	Hevea brasiliensis

[a] Additional information on most of these species may be found in Burkill (1966).
[b] Grown predominantly by Chinese.
[c] Numerous hybrids.

The Malay diet is based on rice. The second most important calorie source, and also an important protein source, is coconut. Coconut cream pressed from grated coconut soaked in water is the base for most curries. Like many southeast Asian languages, Malay has a word, *lauk*, for food eaten with rice (normally as a topping); curries and small fried fish or similar fried relishes are common lauk. The major animal protein source is marine fish, with chickens and eggs second and meat (goat or beef, mostly water buffalo beef) a distant third. The preferred fish are oily, calorie-rich species such as Spanish mackerel (*Scomberomorus spp.*). Curry spices are used in enough quantities to the considered as nutrient contributors to the Malay diet. For instance, chiles, the major spice, are rich in vitamin A, the B complex, and vitamin C (if they are, as usual, fresh and not cooked too long). Cumin, fenugreek, anise, cinnamon, cloves, black pepper and coriander seed are commonly used and may also provide some nutrients (See Table II for details). Most of these spices are imported from India. Vegetables are less important; not only are they used in small quantities, but those used—onions, eggplants, manioc and okra, for instance—tend not to be very valuable vitamin sources. Potatoes, tomatoes and various wild greens are better vitamin sources, though not used in large amounts. (For more detailed discussion of the Malay diet, see Wilson 1970).

Snacks are important in the Malay food system, and these were traditionally either local fruits or other home-raised items. Many of these were good sources of nutrients (Table III). Today, snacks are increasingly likely to be candy, cookies and other items of limited nutritional worth. A vast array of small cakes and candy-like confections is made. Traditionally raw sugar and coconut were the bases of these, but now they tend to be mostly white sugar held together with milled rice—especially glutinous rice.

Traditional food resources were diverse and adequate if the family could acquire the spices and fish. In interior Malaysia many cannot, and deficiencies of iron and vitamin A are common (Wilson personal communication), but in coastal Penang there was little evidence of such deficiencies. A problem does arise among young children who have difficulty accepting the curries (due to high quantity of chiles) and thus have little or no access to many nutritious adult foods. This problem, however, was less severe in traditional times, when children snacked continually on fruit, than it is now that white-sugar products have come to dominate the snack roster in many villages. Children begin to eat curries at about five to seven years of age, but the amounts

TABLE II

Energy value and nutrient content[a] of selected spices[b] and foods consumed in Penang

	Calories (100 g)	(kJ)	Protein (g)	Calcium (mg)	Iron (mg)	Carotene Equiv. Uts Vitamin A	Thiamine (mg)	Riboflavin (mg)	Niacin (mg)	Ascorbic Acid (mg)
Anise seed[c]	415	1736	19.0	693	34.8	d	d	d	d	0
Chile, dried[e]	288	1205	11.7	223	45.0	27,430	1.14	1.53	19.8	184
Cumin seed[c]	332	1389	16.6	1,365	24.3	520	d	d	d	0
Ginger (dried)	301	1259	7.6	180	d	120	.16	.27	8.4	0
Mustard seed	469	1962	26.4	410	20.9	630	.40	.31	7.3	0
Black pepper	325	1359	12.2	d	d	d	d	d	d	0
Turmeric (dried)	337	1410	6.3	d	d	10	d	d	d	0
For comparison:										
Soybeans (whole dry bean edible portion)	400	1674	35.1	226	8.5	10	.66	.22	2.2	0
Rice[f] (milled, polished, unenriched)	366	1531	6.4	24	1.9	0	.10	.05	2.1	0

[a] The relatively high food value of spices is not widely recognized, and data on spices are not easy to find in the United States. The following determinations are from *Food Composition Table for Use in East Asia*, compiled by Woot-Tsuen Wu Leung, Ritva Butrum and Flora Chang (1972).

[b] A typical adult Malay may consume up to an ounce (28.3 g) of spices a day, not counting chiles, and from one to several ounces of chiles. Usual consumption figures are lower but still not insignificant. In a Malay context, a typical curry uses about an ounce of spices and an ounce of dried chiles for a family of about six people. Individuals may partake of anywhere from one to six curries or helpings of curry in a day, the high figure being for special occasions.

[c] Unpublished data available show other spices ranging between the extremes of cumin and anise on the "good" side and turmeric on the "poor." Other seeds of the carrot family (Apiaceae) are especially close to their relatives cumin and anise.

[d] Undetermined.

[e] Fresh chiles are higher in vitamin C, up to 103 mg/100 g with only 10% of the calories of dried chiles.

[f] Rice is tabulated in this processed form because that is the way it is generally consumed in Penang.

TABLE III

Energy value and nutrient content of common fruits regularly and widely eaten in Penang[a]

Tropical Fruits[b]	Calories (100 g)	(kJ)	Protein (g)	Calcium (mg)	Iron (mg)	Carotene Equiv. Uts Vitamin A	Thiamine (mg)	Riboflavin (mg)	Niacin (mg)	Ascorbic Acid (mg)
Banana	100	418	1.2	12	0.8	225	.03	.04	0.6	14
Cashew fruit (not the nut)	53	223	0.8	7	0.6	50	.02	.01	0.5	198
Durian, whole[c]	31	130	0.6	5	0.2	trace	.07	.07	0.3	9
Guava, common	56	234	1.0	15	0.7	75	.05	.04	1.1	132
Jackfruit	94	393	1.7	27	0.6	235	.09	.11	0.7	9
Mango, ripe	62	259	0.6	10	0.3	1,880	.06	.05	0.6	36
Rambutan	64	268	1.0	20	1.9	0	.01	.06	0.4	53
Soursop	59	245	1.0	14	0.5	trace	.08	.10	1.3	24
Temperate-zone Fruits										
Apple	51	213	0.4	10	0.5	20	.02	.03	0.2	4
Orange	40	167	0.8	21	0.3	150	.07	.04	0.4	43
Peach, yellow flesh	43	180	0.8	9	1.0	245	.03	.07	0.4	6

[a] See Table II for reference.
[b] The greater nutritional value of the tropical fruits, especially vitamins A and C, compared to the temperate-zone fruits is striking. Note especially vitamin A for mango and C for cashew fruit (which is actually a swollen stem holding the nut) and guava.
[c] Since ¾ of the whole fruit is waste, multiply figures by 4 to get approximate value of edible portion. Other figures in this table are for edible portion only.

consumed are small until that are about ten years old. Curries are often thought to be too heating in terms of humoral madicine. On the other hand, curries are associated with adult status and with festive occasions, so children want to develop the ability to eat them freely.

Of the Malay farm crop, little is sold. Snacks made at home and either retailed or sold to local shops for resale make up a common source of pin money for women. Vegetables, fruits and betel nuts are sold in small quantities the same way, namely by direct vending or sale to the small local shops that are universal in rural Malaysia. The main sources of cash on the Malay farms, however, are rubber-tapping and the sale of labor power—mostly by men who leave the farm for short or long periods. Much of their income is used for supplementary food—buying more rice, sugar, spices and minor luxury items for feasts. The farms are thus of a pattern familiar in Third World nations, in which subsistence agriculture is used to support laborers (who thus can potentially be paid less than subsistence wages), and the laborers' wages in turn tide the farms over difficult periods.

THE CHINESE FOOD SYSTEM

The Chinese farms of Penang are very different. They specialize in production of the market, in contrast to the Malays, who are more or less subsistence farmers although they sell a certain amount of their produce. The Chinese do eat a good deal of what they raise, however, and many families who do not live by farming have substantial vegetable and fruit plantings. The difference is thus not wholly due to market orientation.

A typical Chinese farm will have pigs, chickens, ducks, and sometimes other animals, and some of the same fruits as the Malay farms—especially small, fast-growing items also familiar in south China, like bananas and guavas. Little or no rice is grown and manioc is not often important. Emphasis is on intensive cultivation of vegetables. The major ones, in addition to those grown in Malay villages, are sweet potatoes, yard-long beans (*Vigna sinensis*, grown as a very long green bean), peanuts, taro, squashes and gourds of several species, and a number of greens including *Amaranthus* sp. and heat-tolerant forms of Chinese cabbage or mustardgreen (*Brassica chinensis*). Spices are far less important in most gardens, with the notable exception of ginger which is much more important among Chinese than Malays. Chiles are much less widely used but are often cultivated for sale. When tree crops are important, they are often grown in orchards—that is, spaced stands of only one or two species of tree—rather than as a mixed economic grove as is seen around Malay houses. The Chinese rigorously resist putting their houses on piles and like to keep bare yards and paths, thus limiting plant and root space. Vegetable fields take up many acres and are intensively cultivated.

This specialization on vegetables and livestock is possible only if a high level of additional inputs is maintained. In particular, an enormous amount of labor is needed, and it must be quite skilled and available at almost all times. Vegetables need frequent weeding and—even in rainy Malaysia—watering. Preparation of the soil for vegetables is difficult to do at all and still more difficult to do correctly. The soil must be well-drained, necessitating planting in raised ridges for most crops. Great effort is expended maintaining the ridges and drainage canals. Constant attention is paid to the beds, which are sprinkled if the day is rainless, sprayed with pesticides and heavily fertilized with both compost and commercial fertilizer. Livestock needs considerable attention, particularly for large-scale market operations. Pigs are not allowed to forage for themselves, but are cared for regularly and specially fed. Pig food consists of any vegetable and other edible waste material, but some pig food must be cooked or specially mixed. Commerical feed is used in increasing amounts. The labor force consists of family members when possible; otherwise local workers are hired.

In addition to labor, fertilizer is used heavily, consisting of both farm manure or compost, and commercial preparations. Insecticides and other agricultural chemicals are regularly used. Vegetable seed must be bought and is not cheap, as prices go, in rural Penang. The poor drainage, poor quality and rapid deterioration after clearing of soils in Penang necessitate constant reworking and addition of compost and fertilizer. However, if all this is done, the land gives very high yields of the finest-quality commercially grown vegetables. (For full accounts of Chinese vegetable growing in the Malaysia-Singapore region, see Clarkson 1968 and Fan 1969, where cost, yield and other figures are detailed.)

The Chinese diet, like the Malay, is based on rice, but here the resemblance ends. In the first place, wheat is important; noodles and bread provide about half as many calories as rice. Sugar is at least as important as wheat. Yet another major calorie source is one totally closed to the Malays: the pig. The Malays, being devout Muslims, avoid all contact with this animal. The Chinese not only eat large quantities of port but use lard as the major cooking fat and thus a major component in virtually every meal or substantial snack. Fish are as important to the Chinese as they are to the Malays, but they are not the main animal protein source. The Chinese prefer delicate, white-fleshed, low-calorie fish and detest the oily, strong-tasting fish savored by the Malays. The Chinese eat large quantities of vegetables, not only those they raise buy also cool-climate varieties (which are easily and cheaply available in Penang because they are raised by the Karo Batak in the highlands of North Sumatra, just across a strait). These exotic vegetables include cauliflower, cabbage, carrots and potatoes. Spices are relatively little used, though a few Malay-influenced dishes make use of them. The standard Chinese meal, corresponding to the Malay rice and curry, is rice with side dishes of vegetables stir-fried in lard with small amounts of finely sliced pork. Noodles with vegetables and pork are a common snack. Whether in soup or fried, they are usually heavily larded. (It should be mentioned that lard is especially typical of the Hokkien and their linguistic relatives, who are the major but not the only Chinese groups in Penang. Other Chinese ethnic groups make more use of vegetable oils.) The Chinese, like the Malays, snack often on fruit; but even more than the Malays, they have replaced fruit with candy and cookies. When I asked why, I was told, "Candy and cookies are made in factories; fruit is raised in *peasant villages!*" The ending of the sentence was spoken with utter scorn. More detailed accounts of the Penang Chinese diet are available (Anderson and Anderson 1972, 1978), and a general overview of the nutritional value of the diet is given here.

Like the Malay diet, it is adequate in all respects if the eater/preparer is reasonably well-off and following the traditional pattern. The greater importance of vegetables and pork compensates for the insignificance of spices and the rather lesser use of fish. The vegetables are much more apt to be those high in vitamins A and C than those in Malay curries. For example, Chinese cabbage has 1200 units of vitamin A equivalent and 40 mg ascorbic acid per 100 g; this and other cabbages of similar value are common in Chinese cuisine but rare in Malay. The reverse is true of eggplant, which has 50 units vitamin A and 6 mg ascorbic acid per 100 g. It is both common and typical of Malay vegetable cuisine (figures from Wu Leung, Butrum and Chang 1972).

However, the large amounts of lard, white flour and sugar are at best of dubious value, and it is these which are increased as diet becomes modernized or as incomes fall. (It should be noted that very few Penang Chinese are short of calories; on the contrary, obesity is common and widespread, even among the relatively less affluent.) The well-known tendency in tropical countries for sweets, highly processed foods, and infant formulas and prepared foods to replace local fruits, vegetables and breastfeeding is extremely advanced among Penang Chinese. This has contributed to severe tooth decay; it was common among children for both deciduous teeth and permanent teeth to decay as fast as they grew in. Other problems of poor nutrition, especially among children, are evident from this acceptance of modern foods. Though the Chinese diet is less spicy than the Malay, it is not lacking in chiles; also, Chinese traditional food beliefs as found in Penang held that most adult foods were too much for the youthful system. These adult foods are said to be "too heating" or "too hard to digest." Thus children lived almost entirely on rice, noodles and sweets (Anderson and Anderson 1972, 1978). It should be remembered, however, that this is a recent development. Use of such snacks has increased steadily during the 20th century, but become much commoner in the last 20 years. The traditional diet of rice (less milled than it is now), pork, fish and large amounts of vegetables and fruits was an excellent one in terms of providing adequate nutrition at low cost.

FOOD SYSTEM DIFFERENCES

The differences between the Malay and the Chinese food systems are based, ultimately, on the differences in demand. Malays use more spice, more fruit, more coconut, different meats and fish. Chinese use more vegetables and, of course, pork. Another aspect of demand is that the Chinese of Penang are mostly urban and fairly secure in livelihood, with a portion of this population relatively affluent. Therefore, they buy most of their food and are able to purchase a high quantity and diversity of foods. The Malays are primarily rural and less well off financially; they produce much or most of what they eat. Therefore, the typical Malay is a true peasant, producing much of his or her own subsistence and marketing only a small amount of goods, commonly nonfood products such as rubber or betel-nut. The Chinese farmer is usually a market gardener. He or she eats the less saleable produce but does not depend on it for survival; instead he/she uses much of the cash received to buy rice and wheat and sugar for bulk calories.

These differences in orientation are maintained by a complex series of political and social factors. The Malays have several crucial disadvantages vis-a-vis the Chinese in market-oriented agriculture, but advantages in regard to nonfood crops. Malays will thus intensify non-food cropping (mostly rubber) if they want cash from their farms or, more likely, simply turn to fishing. Their expertise centers on the production of rice and fruit. Rice is no longer very significant in acreage in Penang. Fruit commands a low price and is being replaced by the preferred, factory-made (or hand-made) sweet snacks. Able-bodied men of Malay farms, those who might otherwise provide the labor and entrepreneurship for more ambitious, market-oriented farming, are often absent for varying periods to take advantage of better economic options.

Due to political tension between Chinese and Malays and consequent preference by the Malay-dominated government in hiring and placing, Malays are able to join the army, police and government service in large numbers. This classic route for upward mobility for rural folk is much less open to Chinese. Young Malay men can also take advantage of the Federal Land Development Authority's schemes to open up less settled land in the interior of the main peninsula of Malaysia. Failing all else, they often turn to commercial small-scale fishing for short periods. Thus the family labor pool of the typical Malay farm is depleted in precisely those categories of manpower and entrepreneurial talent that it can least afford to spare. The cash they remit is not easily invested in the farm in the absence of leaders and able-bodied adult workers; it often is used to provide feasts, help indigent kin and otherwise maintain social status.

Additionally, there is the problem of learning what Chinese entrepreneurs already know about fertilizing, drainage, marketing, accounting and other technical and economic aspects of farming. Last, and perhaps most crucial of all, there is the fact that a Malay who has plenty of food or any reasonable sum of money is expected to share it freely with relatives of whom he is certain to have many (most of them poor), given the interlocking kinship networks and general rural poverty of Penang's Malay community. The Chinese, by contrast, will help impoverished relatives, but with two important differences. They are not expected to help to the point where it hurts their own enterprises severely; and the relatives are expected to repay, somehow, someday. The Malays regard this behavior as inhuman and ostracise anyone in their community who shows signs of it, thus making entrepreneurship difficult at best. This does not mean that the Malays have no entrepreneurial abilities or interests. Everywhere in Malaysia that the Chinese do not dominate the small retail sector, Malays are found—for instance, in the more isolated Malay villages of Penang. But the Chinese are generally more successful.

Another problem for Malay farmers is acquiring the capital and the knowledge for entering market gardening. Capital formation is difficult at best, but given social demands on funds and the tendency to spend any funds that do accumulate in other ways (moving to FLDA land schemes, buying fishing boats, etc.), relatively little is invested in farming, mostly in new tree-crop plantings (especially rubber trees), in upgrading rice cultivation (for example obtaining a small mechanical cultivator) or elsewhere in the traditional subsistence and nonfood farming sector. Rice-growing and tree-cropping have their own specialized knowledge, quite different from that needed to run a vegetable farm. Few sources of useful information were available as of 1970–71 for assisting Malay farmers in dealing with the agricultural and accounting techniques necessary to run a market-oriented operation.

The Chinese, on the other hand, perceive different opportunities. Growing their familiar vegetables for a very large and relatively well-off urban Chinese population, they are sure of a ready market (though with many competitors seeking to satisfy it) and of good prices for what they can sell. This enables them to buy fairly expensive inputs and also removes much of the temptation felt by the Malays to seek better jobs outside of farming. Unemployment among urban Chinese was very high in Penang in 1970–71. Careers in the army and police and many branches of government service were reserved for the Malays, and such hiring prejudice was not confined to the governments. The fishery was already overmanned. Thus sons stayed on the farm, ensuring a large labor force and continuity of operation and knowledge.

Moreover, the Chinese are expert at creating and maintaining personalistic links to wholesalers at all levels. Kinship, language, area from which one's ancestors emigrated, common school experience, political ties and friendship are all rich sources of network ties that can be used to facilitate economic dealings. The incredible ramifications of such networks, and the skill of most persons in manipulating them, are actually the most critical factors in the well-known success of

Chinese businessmen in southeast Asia. (For excellent discussions of how such networks operate, see Skinner 1957, 1958; and Young 1974.) Farmers participate in these to a degree that would surprise anyone who believes the time-honored stereotypes of peasants in less developed countries.

The Chinese emigrated to Malaysia with a considerable body of market gardening skills and have learning and adapted to the tropical environment over more than a century. They are adept at using the somewhat limited agricultural education and extension services available in Malaysia, and also at learning by word of mouth.

All of the above factors, however, would not prevent each group learning slowly from the other—borrowing crops, techniques and skill—if there were not a powerful further factor keeping the two groups separate. This factor is the virulent communal hatred between Malays and Chinese. This ethnic polarization is primarily the result of British divide-and-rule policy, or perhaps . the more general phenomenon of British ethnic stereotyping in the colonial era when colonial administrators and planners rather routinely assumed that every ethnic group had its niche. The stereotype in Malaysia was that the Chinese were born shopkeepers and economic entrepreneurs, while the Malays that left the farm were recruited for the army and police, as is true today. Chinese were allowed or encouraged to become dominant in the marketing, retailing, wholesaling, market-gardening and, in general, the whole entrepreneurial side of the economy. Apparently some conflict between the two groups existed even before the British, but certainly the present animosity is basically a product of the colonial situation. Direct encouragement of separation was important, but probably less so than the indirect encouragement provided by the economic streaming described above. (For fuller accounts of the problem as it evolved and exists today in rural Malaysia, see Anderson and Anderson 1978; Strauch 1981.) Religion is also a component and one that directly affects farming since it interdicts to the Malays that key animal—the pig—which is both an exceedingly important money-spinner and a valuable fertilizer source for the Chinese. As of 1970–71 ethnic polarization had reached a state in Penang where most people feared out-and-out civil war. Riots in Penang in 1968 preceded the larger ones in Selangor in 1969; many Chinese had been killed by Malays, acting in Selangor at least with the direct backing and support of highly placed government officials. In 1970 Mahathir bin Mohamad, today the Prime Minister of the nation, published a book, *The Malay Dilemma*, which advocated a policy of "the Chinese must go" (to borrow the old American racist phrase) in such militant terms that it was banned (Mohamad 1970). Since it was freely sold in Singapore, it was easily smuggled into Malaysia, and banning only served to make it more popular. In the 1970s there was increasingly severe discriminatory legislation aimed at Chinese and their businesses.

In such a situation, hatred and prejudice prevented mutual borrowing. Indeed, Chinese known to me were rapidly abandoning the behavioral strategies they had previously borrowed from the Malays (use of spices in cooking, for instance), and Malays were self-consciously retreating from both western and Chinese modes of life, especially the latter, into a narrow Muslim Malay cultural universe. To say that a food or farming method was identified with the other ethnic group was to give a full and adequate explanation of why it was shunned. This kept the mosaic of rural Penang clearly and sharply defined. In neighboring countries, Thailand and Indonesia, where similar contrasts in farming systems originally existed, but where ethnic rivalry has been neither so long-standing nor virulent, there has been much more accommodation. A blending of traditions is evident. These countries provide a useful contrast, showing that accommodation is not only possible but reasonable, and will occur even in spite of considerable hostility. Only the implacable prejudice found in Malaysia is sufficient to keep borrowing down to the minimal level observed in Penang.

The two food systems potentially have quite different ecological effects. The Malay system is the product of thousands of years of evolution *in situ*, and is adjusted spatially and temporally to the environment. The outstanding development is the mixed economic forest that solidly covers and hides the villages, roads and everything else of humanity as well as the upland landscape in general. Reproducing a tropical forest in its multistoried, dense-canopied, highly diverse structure and flora, it is good enough imitation of the natural vegetation of the area to have similar economic and ecological stability and productiveness—all with very little input from the human farmers. The trees will grow on the infertile soils of Penang—some on the lateritized hills, others on poorly drained sites, and several (coconut, *Terminalia*, rose-apple, and some others) on beach sand. Every area is intensively used. The starch staple, rice, is grown in low-lying floodable areas. Far from depleting the soil, it actually enriches it by trapping washed-in nutrients and by its own decomposition (roots, stalks, etc.) and, in the more general sense in which rice agriculture is the unit of discourse, because of the nitrogen fixed by blue-green algae which grow in the irrigation water. Traditionally, fish and prawns flourished in these paddies and produced additional food, but modern quick-maturing rices do not leave much time for these to grow.

The Malay system, mimicking the diversity and structure of the tropical rainforest, provides subsistence within the context of the food system. Many of the spices and vegetables are perennial crops, and their beds are protected by the canopy layer. For market farming, however, this is not an ideal system. Harvesting is made difficult by the scattering of individual species and the tendency of items to ripen in slow sequence rather than all at once. In any case, the lack of a good market for the foods produced rules it out as a viable economic strategy in Penang. Even marketable items, when produced, are grown in too small quantities and too widely scattered plots to make it economical to assemble them for sale. Small amounts of a wide variety of things, and small but steady yields of ripe items, are just what is needed for subsistence, however. In the tropical climate, it is much easier to manage this way than to try to store large quantities of foods.

The Chinese system, by contrast, is designed to make marketing easier rather than to produce stable yields over a long time. Beds are large and monocropping is common. The widely spaced rows and raised baulks create ideal vegetable-raising conditions but leave the fields open to the full effects of sun and rain. Soil destruction is remedied by application of large quantities of fertilizer. The long-term effects of this attempted recompense on soil structure and erosion are not well known. Pest populations build up on the large fields of succulent foods, and insecticides and herbicides used on them pollute the environment ultimately degrading or destroying its productive capability. The fishery, a necessary and well-integrated part of the traditional local system, has been destroyed throughout inshore Penang state (except very locally) by pollution and overfishing. Much of the pollution is traceable to modernized agriculture, with the plantation sector a worse offender than either of the two food-producing systems discussed here. However, the Chinese small farming system does yield a good cash return quickly in the short run and does not degrade the environment fast enough to allow degradation to serve as a disincentive to the farmers. The Malay food system is adjusted to a longer environmental timeframe; the Chinese is a short-run system. Moreover, based as it is on intensive vegetable and stock rearing, the Chinese system would be convertible into a subsistence agricultural system only with difficulty.

DISCUSSION: A SUGGESTION FOR THE FUTURE

Each system, then, is a solution to a particular problem. Communal hostility prevents merging or fusion of the two. However, as answers to modern Malaysian needs, both have their problems. The Malay system produces too little marketable food for the cities; the Chinese leads to environmental degradation that reduces the productive potential of the region.

Without interethnic conflict, a more satisfactory system could be worked out. Clearly, it would be based on the Malay system, with rice in the low places and tree crops elsewhere as the basic components. More attention would be paid, however, to the economics of tree-cropping. Large stands of coconut-banana intercrop would be maintained.[3] Mixed fruit orchards—durian, rambutan, mangosteen, citrus, guava and the like—would stretch over the hills. The beach also would be a tree-intercropped zone. All of this presupposes that some change of taste, or more accurately change of prestige-ranking, would come over the citizens of Malaysia, such that they would eat fruit when possible and would work actively to develop an export market for their incomparable but little-known produce. At present, mention of such things is met, not only with the scorn of the townsman for the peasant, but also with a refusal to take seriously any food except rice. Rice *is* food, in both Malay and Chinese; *lauk* and *sung* are but flavorings, even to those who have learned western concepts of nutrition and economics. Spice cultivation could be greatly extended and rationalized; again this awaits an expansion of demand.

Vegetables would be integrated into the food system by small plots of the most nutritious ones that yield well in the area, including Chinese greens (cabbage species and amaranths, *Ipomoea*, etc.), tomatoes, peanuts, sweet potatoes, peppers and others. If manioc and taro continue to be grown, much more should be done with their highly nutritious leaves. Bean crops and bean foods could be extensively introduced; the yard-long bean is grown and a few soy products used, but more attention to these, plus introduction of mung beans, winged beans (*Psophocarpus*) and other legumes is desirable.

Intensive livestock rearing in an ecologically integrated context has been tried in Penang. On one large farm near my field site, pig farming (in modern piggeries) was combined with fish farming. The fish ponds were fertilized from the sties and the water runoff from them used to irrigate—and fertilize—vegetables. Tree crops, also thus fertilized, were grown on slopes and areas of well-drained soil. This project failed because of distance from markets and problems (as usual) of marketing the more traditional fruits and other foods, as well as because of the problems common to all pioneers—locating dealers, debugging the system, etc. It showed what could be done. The pig is probably the most successful domestic animal in the area, except for poultry, but is of course unacceptable to Muslims. Goats afford a ready substitute; they flourish on the coarse local feed and produce milk into the bargain. In spite of efforts directed toward developing a livestock industry based on goats and water buffaloes, no plans have succeeded. Here the prime reason is simply that no one has bothered to try seriously.

Meanwhile, the diet of the region is rapidly and inexorably modernizing. White (and often unenriched) rice and flour, white sugar, processed and often highly saturated fats and oils and other highly processed foods are becoming more and more available and widely used. Condensed milk and infant foods are replacing breast-feeding. The ruin of the fishery has left poorer people without any readily available animal protein. Traditional fruits and vegetables are being abandoned. So far, the detrimental effects of these changes have been balanced out for most people by the fact that increasing affluence and still somewhat conservative tastes have combined to make people buy vegetables and spices and (when possible) meat in higher proportions than they used to. However, the poor are already in some nutritional trouble. (Deficiencies of iron and vitamin A are reported from poorer Malay communities in nearby states.) Obesity, heart disease, dental caries and diabetes are all increasing rapidly in incidence.

CONCLUSION

The conclusion that has most theoretical significance is that narrowly focused economic considerations are not always adequate to pre-dict ecological and nutritional choices, but we need not retreat to fuzzy and idealistic notions of culture or blind tradition and values either. In the present case, two ideological factors, communal hate

E

and the enormous prestige of anything modern and factory-made, serve as devastating constraints on real progress and development (defining these as more or better resources *per capita*). Originally these forces may have had economic rationales; communal tension was certainly exploited and furthered by the British and by many post-British politicians for "divide and rule" reasons. Westernization is of economic value in that people who act, dress, talk and eat like westerners (or westernized Asians) can parlay these matters of appearance into better jobs and better entrepreneurial opportunities; but here the prestige egg comes before the economic chicken, for it is the prestige of the modern that creates the advantage (Anderson and Anderson 1978; Strauch 1981; Syed H. Al-Atas 1971, unpublished observations). People may in the long run be rational about material calculations, but in the short run it is often such failures as these with which we are concerned. Even the successes may often be more emotion-driven than we suspect. A long list of studies in economics and cultural ecology could be compiled to show that people are broadly rational about material ends and that highly abstract terms like values and traditions tell us little. I argue that one must take account of yet another aspect: the intense emotions that sway people in their daily lives. When these become shared by whole populations and intensely felt, they can be powerful forces, influencing ecological adjustments and all other parts of life.

NOTES

1. An earlier version of this paper was read at the Southwestern Anthropological Association meetings in 1978. I am very grateful to Dr. Dorothy Cattle for valuable and extensive editorial suggestions. The research on which this paper was based was supported by the National Science Foundation; I am deeply grateful to them and to my many friends, helpers and informants in southern Penang Island and elsewhere in Malaysia. The University of California, Riverside, provided support for analysis of data. For a full account of the research, including methodology, see Anderson and Anderson 1978.
2. In the academic and research world, these two realms of study are not segregated by department and publication outlet, they are even segregated by sex. Agricultural sciences is perhaps the most totally male-dominated of all academic fields. Nutrition and food consumption is almost equally dominated by females in point of numbers, but the few male workers in this field are highly visible, which in itself gives some insight into academic sexism.
3. This has been tried in Penang but failed due in part to the opportunities for disease contagion that large stands afford; this is a perpetual problem for tropical agriculture, of course. Small stands are uneconomic; large stands attract epidemics.

REFERENCES

Anderson, Eugene N., Jr. 1972. The life and culture of Ecotopia. pp. 264–283 in *Reinventing Anthropology*, ed. by Dell Hymes. New York: Random House
Anderson, Eugene N., Jr. and Marja L. Anderson. 1972. Penang Hokkien Ethnohoptology. *Ethnos* 1–4:134–147
____1978. *Fishing in Troubled Waters*. Taipei: Orient Cultural Service
Child, Reginald. 1964. *Coconuts*. London: Longmans, Green.
Clarkson, James. 1968. *The Cultural Ecology of a Chinese Village*. Dept. of Geography. Chicago: University of Chicago.
Diener, Paul, Donald Nonini and Eugene Robkin. 1978. The dialectics of the sacred cow: Ecological adaptation vs. political appropriation in the origins of India's cattle complex. *Dialectical Anthropology* 3:221–241
Diener, Paul and Eugene Robkin. 1978. Ecology, evolution, and the search for cultural origins: The question of Islamic pig prohibition. *Current Anthropology* 19:493–540
Fan Shuh Ching (Ed.). 1969. *Farming in Singapore*. Kuala Lumpur: Union Cultural Organization
Harris, Marvin. 1966. The cultural ecology of India's sacred cattle. *Current Anthropology* 7:51–66
____1968. *The Rise of Anthropological Theory*. New York: Crowell
____1979. *Cultural Materialism*. New York: Random House
McCay, Bonnie J. 1978. Systems ecology, people ecology, and the anthropology of fishing communities. *Human Ecology* 6:397–422
Mohamad, Mahathir bin. 1970. *The Malay Dilemma*. Singapore: Donald Moore
Skinner, G. William. 1957. *Chinese Society in Thailand*. Ithaca, N.Y.: Cornell University Press
____1958. *Leadership and Power in the Chinese Community of Thailand*. Ithaca, N.Y.: Cornell University Press
Strauch, Judith. 1981. *Chinese Village Politics in the Malaysian State*. Cambridge, Mass.: Harvard University Press
Vayda, Andrew and Bonnie I McCay. 1975. New directions in ecology and ecological anthropology. *Annual Review of Anthropology* 4:293–206
Wilson, Christine. 1970. *Food Beliefs and Practices of Malay Fishermen*. Ph.D. dissertation, Dept. of Nutritional Sciences, Univ. of California, Berkeley.
Wu Leung, Woot-Tsuen, Ritva Butrum and Flora Chang. 1972. *Proximate Composition, Mineral and Vitamin Contents of East Asian Foods. Part 1: Food Composition Table for Use in East Asia*. Altanta: U.S. Dept. of Health, Education and Welfare. Rome: U.N. Food and Agriculture Organisation
Young, John. 1974. *Business and Sentiment in a Chinese Market Town*. Taipei: Orient Cultural Service

ADDITIONAL READING

Allen, Bety Molesworth. 1967. *Malayan Fruits*. Singapore: Donald Moore
Burkill, Isaac H. 1966. *Dictionary of the Economic Products of the Malay Peninusla*. 2nd edition (orig. 1935). Kuala Lumpur: Ministry of Agriculture and Cooperatives
Consumers' Association of Penang. 1976. *Pollution: Kuala Juru's Battle for Survival*. Penang: CAP
Firth, Raymond. 1966. *Malay Fishermen: Their Peasant Economy*. 2nd edition (orig. 1946). London: Routledge and Kegan Paul
Firth, Rosemary. 1966. *Housekeeping among Malay Peasants*. London: London School of Economics

Fox, James J. 1977. *Harvest of the Palm*. Cambridge, Mass.: Harvard University Press
Geertz, Clifford. 1963. *Agricultural Involution*. Berkeley: University of California Press
Ho, Robert. 1967. *Farmers of Central Malaya*. Singapore: Donald Moore

CHAPTER 6

NUTRITIONAL CONSEQUENCES OF THE TRANSFORMATION FROM SUBSISTENCE TO COMMERCIAL AGRICULTURE IN TABASCO, MEXICO[1]

KATHRYN G. DEWEY
Department of Nutrition
University of California
Davis, California 95616

ABSTRACT

Recent changes in the area of the Plan Chontalpa in Tabasco have greatly reduced the production of subsistence crops by rural families, resulting in decreased crop diversity and a concomitant increase in the degree of dependence on outside sources of food. Results from a nutrition survey of 149 families demonstrate that dietary diversity, dietary quality and nutritional status of preschool children are negatively associated with lower crop diversity and increased dependence on purchased foods. Dietary deterioration is illustrated by the negative relationship found between nutritional status and increased sugar consumption. The assumption that a rise in income accompanying the adoption of commercial production will automatically lead to improved nutrition is challenged: income levels were not found to be consistently related to nutritional status. Children of families who have converted to cattle production, despite greater land availability and family incomes, do not have improved nutritional status. In the study area, where wages are low and food prices are very high, the value of a higher degree of self-sufficiency in food is recognized, yet families continue to switch to cash crops due to the environmental, economic and time constraints imposed by the system of commercial agriculture in which they participate. The solution is not to return to traditional subsistence farming, however, but to

Reprinted, with permission, from *Human Ecology 9(2)*: 151–187, 1981.

determine under what conditions a more progressive form of agricultural change can occur.

INTRODUCTION

In recent reviews of the nutritional impact of agricultural change (Dewey 1979; Fleuret and Fleuret 1980), the consensus has been that for poor families in rural areas, a shift to commercial agriculture may often lead to a decline rather than an improvement in nutritional status due to the fact that the conversion involves changes that are far more complex than the mere substitution of cash crops for subsistence crops. Such a transformation generally produces fundamental alterations in the socioeconomic structure of rural communities, including changes in land tenure relations, the degree of involvement with external markets and the nature of rural employment possibilites. A number of studies have demonstrated that even when food production is increased, agricultural development efforts usually benefit primarily the wealthy upper class, while the majority of rural peasants continue living in an impoverished state (Hernandez et al. 1974; Taussig 1978; Palmer 1974). The shift to commercial agriculture has several consequences for small farmers that may jeopardize their economic success. First, cash cropping usually involves a much greater risk, which increases the possibility of indebtedness; and second, commercial agriculture commonly lessens the control of the small farmer with respect to over-production, thereby increasing his dependency on outside agencies or institutions (Stavenhagen 1978; Pearse 1975; Gudeman 1978). In addition, changes in the focus of production from food to a commodity may alter priorities so that they are no longer in line with human needs (as illustrated by Teitelbaum 1977). Changes in the labor requirements of cash crops also may increase the caloric needs of the laborer, resulting in poorer nutrition for other family members (Gross and Underwood 1971).

Many of the nutritional consequences of agricultural change result from the fact that the goals of subsistence and commercial production are very different. Under the former, food produced is primarily for home consumption, while under the latter, it is produced only for its exchange value (Gudeman 1978). Because of this difference, practices that in the past assured a varied, nutritious diet, such as cultivating a diversity of crops and collecting wild plants and other foods, often give way to a reliance on crops grown in monocultures, with little or no fallow cycle to restore secondary vegetation. The replacement of food with cash may lead to dietary deterioration if cash is available only once or twice a year, if cash is spent on non-food items or if the cash received is inadequate to purchase a nutritionally adequate diet (Dewey 1979). While all of the these consequences have been suggested as important influences on nutrition, very few case studies have given detailed attention to them. This study will examine the relationships between subsistence production, commercial agriculture and nutrition, using a case study from southern Mexico.

The state of Tabasco, Mexico, represents a microcosm of the rapidly changing conditions that are characteristic of much of the developing world. The area has undergone considerable agricultural development in recent years, resulting in a rapid expansion of land devoted to commercial cash crops and pasture for cattle. (See Dewey, 1980a for a detailed description.) In addition, Tabasco is at the center of Mexico's recent expansion of oil exploration and production and has experienced major economic changes as a result. A previous paper (Dewey 1980b) has discussed the impact of a large-scale agricultural development project in Tabasco, the Plan Chontalpa, on the diet and nutrition of families in the area. Here the focus will be more specifically on the nutritional impact of the shift from subsistence to commercial agriculture on a family level. Because there is considerable variation in the extent to which individual families in the study have devoted their own plots to commercial production, it is possible to compare families who are relatively self-sufficient in food with those who are more dependent on purchased foods. The central questions to be discussed are: How do subsistence production, crop diversity and dietary diversity relate to dietary quality and nutritional status? What are the interrelationships between income, subsistence production and nutrition? What factors have led to increased cash cropping on a family level in the Chontalpa area?

THE STUDY AREA

Tabasco is a hot, humid tropical lowland area (rainfall 2000–4000 mm [80–160 inches] per year) located on the Gulf Coast in southeastern Mexico. Because much of the area was subject to periodic flooding for several months each year, an extensive drainage project was begun in the 1950s which was the forerunner for the Plan Chontalpa, initiated

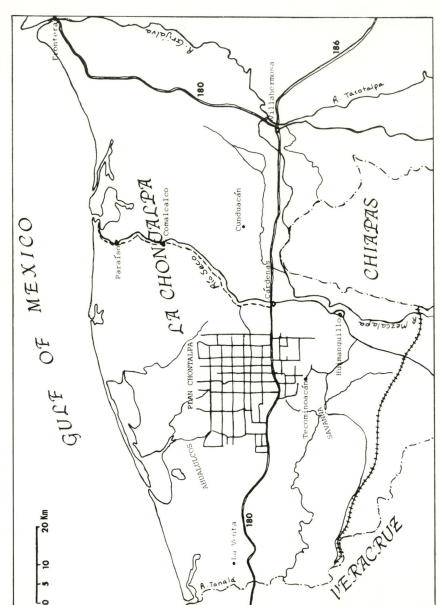

Figure 1. Map of La Chontalpa and environs.

by the Mexican government in the late 1960s in order to increase agricultural production in the area (See Figure 1.). Within the Plan Chontalpa there are 22 adjacent collective *ejidos* (group farms) which were formed through the expropriation of existing ejidal and private lands and the mandatory relocalization of residents into the new collectives. Approximately 6000 families were relocated, nearly all of whom were residents in the area prior to the Plan. Agriculture in each ejido is based on a combination of private and collective production: Each family has a two-hectare (5 acre) plot of land to farm as they wish, in addition to 13 hectares (32 acres) per family devoted to the collective production of cash crops and cattle. Members of each ejido (*socios*) are paid wages for their work on the collective land. Approximately 200 families live in rows of government-constructed cement-block houses in a central village (*poblado*) within each ejido and are provided with basic services such as water, electricity, primary schools and health clinics. Surrounding each family's house is a small *solar* which is generally used for fruit and shade trees and for raising domesticated animals. Although members of the Plan make up the majority of the workers, there is a large number of *libre* (free) or non-member families who attempt to find work within the ejidos as well. Libre families generally live in thatched houses on the outskirts of each poblado, and are usually relatives or children of socios or are families who were not able to become members of the Plan when it was begun (primarily landless laborers).

Outside of the Plan Chontalpa are scattered villages in which subsistence and commercial farming are practiced to varying degrees, usually in combination with wage labor. The village of Tecominoacán (henceforth abbreviated to "Teco"), located on the southern border of the Plan Chontalpa, represents an area in which a mostly self-provisioning subsistence agriculture is being increasingly replaced by cattle production. Teco was chosen for this study in order to investigate the impact of the expansion of cattle production in addition to serving as a comparison with the Plan Chontalpa. About half of the approximately 165 families of Teco have at least part of their land devoted to pasture for cattle.

The traditional agriculture of the Chontalpa area is based on a slash-and-burn form of cultivation using mostly manual labor. The major subsistence crops are maize, beans, manioc, sweet potatoes and squash, cultivated since pre-Conquest times, and plantains and rice, which were introduced from the Old World. Cacao, which is native to the area, has been grown for centuries, not only as a subsistence crop

TABLE I
Traditional Crops of La Chontalpa

Crop	Scientific name	Crop	Scientific name
Staple crops		Fruits	
Maíz (maize)	Zea mays	Melón (cantaloupe)	Cucumis melo vars.
Frijol negro (black bean)	Phaseolus vulgaris	Sandia (watermelon)	Citrullus vulgaris
Frijol pelón (cowpea)	Vigna sinensis	Piña (pineapple)	Ananas sativas
Calabaza (squash)	Cucurbita pepo vars.	Mango (mango)	Mangifera indica
Yuca (manioc)	Manihot esculenta	Naranja (orange)	Citrus auranteum vars.
Camote (sweet potato)	Ipomoea batatas	Limón (lime)	Citrus limonia vars.
Macal (taro)	Colocasia esculenta	Mandarina (mandarin orange)	Citrus reticulata
Malanga (taro)	Xanthosoma sagittifolium	Lima (lemon)	Citrus limon
Arroz (rice)	Oryza sativa	Quihiniquil	Inga quinicuil
Plátano (plantain)	Musa sp. and vars.	Nance	Byrsonima crassifolia
		Papaya (papaya)	Carica papaya
Other crops		Ciruela (plum)	Spondias purpurea
Cacao (cocoa)	Theobroma cacao	Guineo (banana)	Musa sp. and vars.
Tomate (tomato)	Lycopersicon esculentum	Aguacate (avocado)	Persea americana
Chayote	Sechium edule	Chinin	Persea sp.
Pepino (cucumber)	Cucumis sativas	Anona	Annona squamosa
Col (cabbage)	Brassica oleracea	Guanabana	Annona muricata
Cebolla (onion)	Allium cepa	Guayaba (guava)	Psidium guajava
Ajo (garlic)	Allium sativum	Tamarindo	Tamarindus indica
Rábano (radish)	Raphanus sativus	Zapote	Achras zapota
Papa voladora	Dioscorea bulbifera	Chico zapote	Calocarpum mammosum
Ñame	Dioscorea alata	Castaña	Artocarpus communis
Café (coffee)	Coffea arabica	Caimito	Chrysophyllum cainito
Caña de azucar (sugar cane)	Saccharum officinarum	Marañón	Anacardium occidentale
Chile (chile pepper)	Capsicum annuum vars.	Capulín	Muntingia calabura
Greens and spices		Granada	Passiflora edulis
Chaya	Cnidoscolus chayamansa	Coco (coconut)	Cocos nucifera
Chipilín	Crotalaria maypurensis	Guaya	Talisia olivaeformis
Epazote	Chenopodium sp.	Almendra	
Momo		Huspi	
Cilantro (coriander)	Coriandrum sativum	Pitaya	
Achiote	Bixa orellana		
Pimienta (black pepper)	Pimenta officialis		

but as a major export crop. Maize is the most important staple crop and is consumed primarily in the form of *tortillas* and a drink known as *pozol*, made from ground maize and cocoa beans mixed with water. In addition to the major staples, a wide variety of fruits and other crops are grown, often in the area around the house (the solar) (See Table I).

Commercial agriculture is now the dominant form of land use in the Chontalpa area. The trend toward increased production of sugar cane, bananas, rice and especially cattle was greatly accelerated by the Plan Chontalpa, with the result that subsistence production declined dramatically (Dewey 1980b). Accompanying the transformation from subsistence to commercial agriculture is a change in labor structure : What was once a largely self-provisioning peasantry is now almost completely dependent on wage labor, whether families are part of the Plan Chontalpa and must work for wages on the collective land or whether they are outside the Plan and must earn cash through day-labor.

In Teco land structure and subsistence production were drastically altered by the Plan. Approximately half of the land in the village was expropriated when the ejidos of the Plan were formed and the families affected either moved to a newly-formed poblado as socios or remained in Teco and found other land to cultivate. Unfortunately for the villagers, the land that was expropriated was the most fertile alluvial land near the river, while what remains is poorer-quality savanna land. Nearly all of the villagers complain that they cannot produce adequate quantities of staple crops on the land that remains, largely due to the invasion of cropland by grasses from nearby cattle pastures. As a result most of the families in Teco are forced to find work as day-laborers outside of the village.

The recent expansion of oil production in Tabasco has greatly affected the entire study area. While there are increased job opportunities for skilled laborers, the primary effect for most families has been the inflation of food prices in the area rendering the shift away from subsistence production even more detrimental to poor rural families.

THE STUDY DESIGN

In order to assess the influence of agricultural change on diet and nutrition, a nutrition survey of 149 preschool children was conducted, both within and outside the Plan Chontalpa. Nutritional status was evaluated for children aged 2–4 years because children of this age are very susceptible to malnutrition and indices of growth can be easily measured as a reflection of nutritional condition.[2] The study was designed as a cross-sectional comparison of children in three groups: families who are members of the ejidos of the Plan (socios), non-member families (libres) and families in the nearby village of Teco. In the Plan Chontalpa, 5 ejidos were selected using a table of random numbers. Maps were constructed for all 5 ejidos and for Teco, and each house was assigned a number. In each ejido families were selected at random unitl there were 15 socio families with children 2–4 years old. Libre families were selected at random from the three ejidos that had a large number of libres living in them; from the other two all of the libre families were selected, until there were 25 libre families in total. In Teco, 85 families were randomly selected, of which 48 had children 2–4 years old.

All families in the study were interviewed in the home to collect information on family history and demography, income, employment, education, child-care practices, food consumption and agricultural practices and yields. In addition, two 24-hour dietary recalls were completed for each child in the study.[3] All interviews were completed during the summer growing season (June–August, 1978). In the winter, a subsample of 44 families was revisited in order to collect more detailed information on diet, agricultural practices and hours of labor.

Nutritional status of the children was assessed using measures of anthropometry, biochemical measures in the blood and a clinical evaluation. In addition, stool samples for each child were analyzed to determine gastrointestinal parasite levels. All children were measured and examined in one month (September, 1978) with the help of local medical personnel. The subsample of 44 children was remeasured in the winter in order to check for seasonal variation in nutritional status. The general results with respect to nutritional status are discussed elsewhere (Dewey 1980a); this article refers only to the anthropometric measures as they relate to dietary quality. All anthropometric measures (except weight-for-height) are expressed in terms of the percent of the median in the appropriate age and sex category for each child, using reference data for well-nourished children in Mexico (Ramos Galvan 1975). Weight-for-height is expressed as a percentile, using the same reference data.

Because the data[4] from this study come from a cross-sectional survey conducted at only one time, the conclusions drawn depend on a comparison of families who still maintain a certain degree of self-sufficiency in food through subsistence production with those who no longer grow a diversity of crops for home consumption. There are limitations to this type of cross-sectional design, due to the fact that the changes in agriculture in the area have affected *all* families; one cannot, therefore, compare nutritional levels "before" and "after" the introduction of widespread commercial agriculture. Nonetheless, an historical perspective based on information related by families in the study can be used in combination with the cross-sectional comparison to give a more complete understanding of the consequences of agricultural change in the Chontalpa area. Before discussing the results of this comparison, attention will be focussed first on the contrast between subsistence and commercial agriculture and on the degree of cash-cropping practiced in the three major groups of the study: socios, libres and families in Teco.

SUBSISTENCE VERSUS COMMERCIAL PRODUCTION

Because the term subsistence is used with many different connotations, it is necessary to define how it will be used in this article. Gudeman (1978) has described what he calls a subsistence system in which the producer, means of production and product are not separated but are kept within the household-farm unit; thus free labor and a labor market do not exist. When labor is exchanged it is done mainly on a reciprocal basis, and because the goods produced are not sold, there is little accumulation of surplus wealth to be reinvested in future years. This system thus effectively limits the extent to which wealth becomes stratified within a community.

Such a "pure" form of family subsistence production no longer exists on a community level in Tabasco. While individual families may come close to being completely self-sufficient in food, the impact of the market economy in the region has been felt for at least several decades. Indeed, Wasserstrom (1978) points out that even for the relatively isolated indigenous groups of Mexico, no truly traditional subsistence economies have existed since the time of colonialism. The term "subsistence" refers solely to the use of agricultural goods for home consumption and not to an entire subsistence system: It is possible for a family in this study to devote all of its land to subsistence crops and to be relatively self-sufficient in food, for

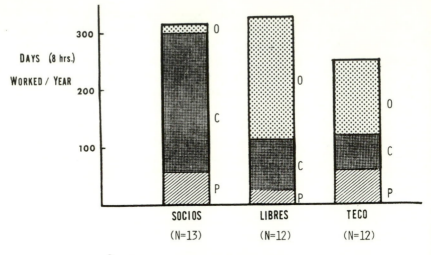

Figure 2.   Wage labor vs. labor in family plot or collective land in Plan Chontalpa and outside Plan Chontalpa.

TABLE II

Measures of mean income (in pesos (1978) per week) of families in two locations in Tabasco

| | PLAN CHONTALPA | | TECOMINOACAN |
	Socios	Libres	
Wages	589 (370–854)[a]	680	574
Crop sales	135 (10–376)	20	21
Sales of fowl and livestock	14 (2–32)	10	108
Share of earnings from collective	157 (16–654)	0	0
Value of subsistence production	112 (60–191)	17	89
Total income + value	1024 (594–1914)	738	803
Per capita income	197 (142–302)	146	143

[a]Range of means in five poblados

example, and yet at the same time earn wages by working outside of the household.

Figure 2 illustrates the average number of days devoted to work on the family plot as compared with wage labor within the Plan Chontalpa or outside of the ejidos. This figure reveals that, for most families in the study, labor on the family plot constitutes only a fraction of total labor output in one year, especially for libre families who generally have little or no land on which to grow crops. Table II lists the average incomes for socios, libres and families in Teco and the components contributing to income (the value of pesos at the time of the study, 1978–79, was 22.7 pesos per U.S. dollar). While these figures should be regarded with great caution, since it is extremely difficult to obtain accurate data on income, it is obvious that in each group the major means of livelihood (on the average) is from wages rather than from agricultural production.

Even if a considerable amount of time is devoted to the family plot, the food produced is often not used solely for home consumption: Excess of basic crops such as corn or beans may be sold to the local store or to merchants who come through the villages in trucks to purchase food for sale elsewhere (*coyoteros*). If a family needs cash, part or all of a crop may be sold even before the family needs are met, obliging them to buy corn or beans at a higher price later. Crops such as squash, manioc and other fruits and vegetables are often sold on a local basis, to neighbors or in the open on the main street of the village. Some socios of the Plan have devoted most or all of their plot to a cash crop such as cacao, bananas or sugarcane. Cacao is sold to the local association of cacao growers, which is responsible for marketing. Bananas are usually sold to coyoteros, and sugarcane is grown through contracts with the local sugarmill.

The extent to which the family plot is used for subsistence vs, commercial (cash crop) production in the Plan and in Teco is shown in Table III. While these figures are based on estimates of the yields of subsistence and cash crops given by the families we interviewed, and therefore are subject to considerable error, they do provide a basis by which to compare families. As shown in the table, socios and families in Teco are similar in terms of the mean value of subsistence production, but socios have a higher mean value of commercial production and a correspondingly higher mean proportion of their family plot devoted to crops which are sold. As expected, the mean values of both subsistence and commercial production are low for libre families. These figures reveal a pattern that we found during the

TABLE III

Family agricultural production in two locations in La Chontalpa, Tabasco

| | PLAN CHONTALPA | | | | | | TECOMINOACAN | | |
| | Socios | | | Libres | | | | | |
	Mean	Range	S.D.	Mean	Range	S.D.	Mean	Range	S.D.
Value of subsistence production[a] (pesos/year)	5831	0–20,000	5573	893	0–6800	1935	4632	0–30,000	6389
Value of commercial production[b] (pesos/year)	7045	0–81,600	14,971	1045	0–23,000	4686	1095[d]	0–22,750	3446
Total value of agricultural production (pesos/year)	12,876	0–81,600	15,682	2091	0–23,000	5224	5774[d]	0–34,750	8146
Percent of land in cash crops[c]	29.1	0–100 N=70	30.7	21.7	0–100 N=12	33.8	8.7[d]	0–50 N=33	13.7

[a] Calculated by determining the amount of money a family would have earned if it had sold the crops used for home consumption.

[b] Value of crops sold, not including value of cattle produced on pasture lands.

[c] Percent of cultivated land used for cash crops (just those families with land in crops during the prior year). If a crop was both sold and used for home consumption, the proportion sold was multiplied by the total land for that crop.

[d] Difference between Socios and Teco significant, Mann-Whitney U test, p < 0.001

(S.D. = standard deviation).

TABLE IV

Self-sufficiency in food in two locations in Tabasco

| | PLAN CHONTALPA | | TECOMINAOCAN | x^2 P-value |
	Socios	Libres		
Degree of dependence on purchased foods[a] (range in parentheses) N =	0.69 (.12–1.0) (76)	0.77 (.40–1.0) (25)	0.67[b] (.12–1.0) (48)	
Percent of families who produce more than half of what they consume for:				
corn	76.3	28.0	39.6	0.0001
beans	43.4	8.0	20.8	0.0034
plantains	40.8	24.0	48.0	0.0300
fruits	17.1	16.0	29.7	0.0563
chicken	31.6	36.0	70.9	0.0000
eggs	18.7	24.0	43.7	0.0075

[a] Calculated using a composite value of the proportion purchased vs. the proportion produced of 11 foods: corn, beans, rice, squash, manioc, vegetables, plantains, fruits, chicken, eggs and pork. A value of 3 was assigned if all of a given food was purchased, 2 if more was purchased than produced, 1 if more was produced than purchased, and 0 if all was produced. The values for the 11 categories were averaged to give an index ranging from 0 to 3, reflecting the degree of dependence on purchased foods. This index was subsequently multiplied by 0.333 to give a more easily interpreted range of 0–1.

[b] Difference between *libres* and others significant, Mann-Whitney U test, P < 0.05.

interviews: More socios are switching to cash crop production than are families in Teco. This may be because most of the land in Teco is no longer suitable for many crops and families who switch to commercial production in Teco usually go into cattle production intstead (as discussed further in a later section).

The degree of self-sufficiency in food was assessed by a measure of the degree of dependence on purchased foods as shown in Table IV. Again, socios and families in Teco are similar for this measure, while libre families are significantly less self-sufficient in food. While socios and families in Teco are similar in their overall dependence on purchased foods, socios produce more of the basic crops such as corn and beans, while families in Teco produce more fruits, chicken and eggs. Families in Teco are unable to produce as much of the staple crops because of the poor quality of the land in their family plots. Those who do grow corn and beans usually do so on a plot of more fertile land that they rent outside of Teco—often from a socio in the ejido of the Plan Chontalpa that was formed from the expropriated alluvial lands of Teco. On the other hand, socios produce less of the items that are produced in the solar—fruits and domesticated animals —because the solares of the poblados in the Plan are smaller, less well-developed and closer together than in Teco.

SUBSISTENCE PRODUCTION, CROP DIVERSITY, DIETARY DIVERSITY AND NUTRITION

Figure 3 outlines two possible routes by which subsistence production may theoretically relate to dietary quality and nutritional status. The left half illustrates the hypothesis that the diversity of crops grown on a family plot will influence the diversity of the family diet: Other investigators have shown (Beaudry-Darisme, Hayes-Blend and Van Veen 1972) that dietary diversity in turn can be imporant in determining dietary quality, which would then influence nutritional status. The right half of the figure suggests that the total value of subsistence production, in combination with the diversity of crops grown, determines the degree of dependence on purchased foods. This may adversely affect dietary quality if purchased foods are of inferior nutritional value or if foods of higher nutritional value, such as animal protein sources, fruits and vegetables, are prohibitvely expensive.

The results found for socio families and families in Teco indicate

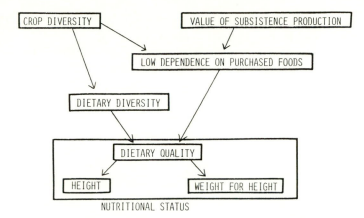

Figure 3. Hypothesized relations between crop diversity, dependence on purchased foods, and nutritional status.

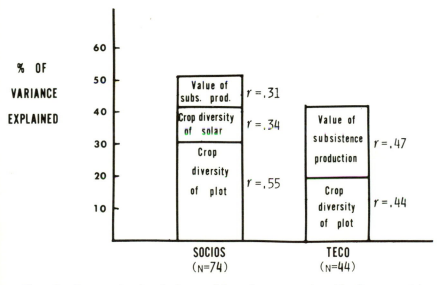

Figure 4. Factors related to the degree of dependence on purchased foods. r = partial correlation coefficient.

that both of the causal chains seem to be important. Libre families were not included in the analysis, as they have so little access to land on which to grow crops and are, therefore, more influenced by non-agricultural factors (discussed in a later section). In the results described below, factors such as income level and other covariables have been controlled as carefully as possible by determining partial correlation coefficients in each analysis.[5]

Figure 4 illustrates the first level shown in Figure 3: the influence of subsistence production and crop diversity on the degree of dependence

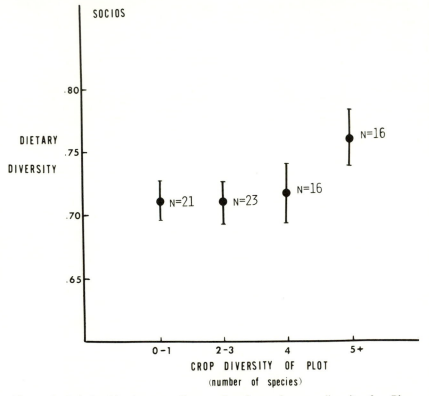

Figure 5. Relationship between dietary diversity and crop diversity for Plan Chontalpa socios (means $+/-$ one standard error). The difference is significant between families with $5+$ crops and all other families (t-test, $P < 0.05$).

Dietary diversity was calculated using an adaptation of a commonly-used index of species diversity (Shannon-Wiener function, Krebs, 1972):

$$H' = \sum_{i=1}^{46} p_i \log p_i$$

where $p_i = \dfrac{\text{calories contributed by food category } i}{\text{total calories consumed in 24-hour recall}}$,

with 46 separate food categories. A mean value was obtained by averaging the diversity indices for the two recalls per child.

on purchased foods. Results are shown separately for socios and families in Teco in Figures 4, 5 and 6, as these two groups appear to be influenced by the two sets of factors of Figure 3 to somewhat different degrees. As shown in Figure 4, 40–50% of the variance in a multiple regression of the degree of dependence on purchased foods is

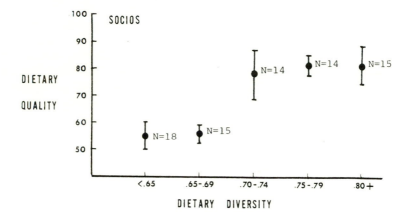

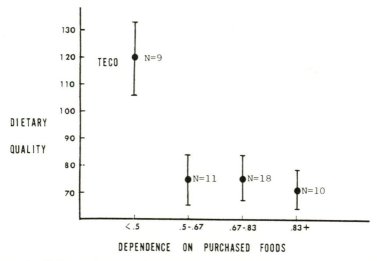

Figure 6.   Factors relating to dietary quality in Mexican Chontalpa communities (means +/− one standard error). For *socios*, dietary quality is related to dietary diversity (Kruskal-Wallis test, p < 0.001). For families in Teco, dietary quality is related to the degree of dependence on purchased foods (Kruskal-Wallis test, p < 0.05).

explained by crop diversity and the value of subsistence production. Income is unrelated to the degree of dependence on purchased foods. The extent to which a family relies on cash for food therefore appears to be determined primarily by the amount and variety of food they grow themselves.

Figure 5 presents the relationship between crop diversity and dietary diversity, using dietary intake data from the two 24-hour recalls per child to calculate an index of dietary diversity. Among socios, children of families with five or more crops in the family plot have a greater diversity of foods in the diet than do children of families with fewer crops. For families in Teco, however, the diversity of the family plot is not related to dietary diversity, perhaps because so few crops are grown on the savanna-type soils of the land that remains to them.

Figure 6 shows the factors most strongly correlated with the next level of Figure 3, dietary quality, defined as the mean percent of the Mexican Recommended Dietary Allowances (Hernandez Chavez and Bourges 1977) consumed of nine nutrients (energy, protein, calcium, iron, thiamine, riboflavin, niacin, ascorbic acid and Vitamin A), as calculated from the dietary recalls. For socios, there is a strong positive relationship between dietary diversity and the quality of the child's diet. It is important to control for the total caloric intake of each child, however, as one would expect both dietary diversity and dietary quality to increase with greater caloric intake. When caloric intake is included in a multiple regression, the partial correlation of dietary quality and dietary diversity is still significant ($r = 0.30$; $P < 0.01$). This result, therefore, supports the argument that dietary diversity can be an important element contributing to good nutrition.

In Teco, dietary diversity is not an important determinant of dietary quality, but there is a *negative* relationship between the degree of dependence on purchased foods and dietary quality. The more a family must buy food, the poorer the quality of the child's diet. This relationship was not found for socio families.

Finally, Figure 7 shows the last level of Figure 3, the relationship between dietary quality and nutritional status, as reflected by the two most useful anthropometric indicators, height (which reflects long-term growth), and weight-for-height (which reflects current nutritional status). As expected, children (of both socios and Teco families) whose recalls indicated a diet of better nutritional quality are taller than children with poorer quality diets (Student t-test, $P < 0.05$). The same is not true for weight-for-height, however. There is no significant difference in weight-for-height between these two groups.

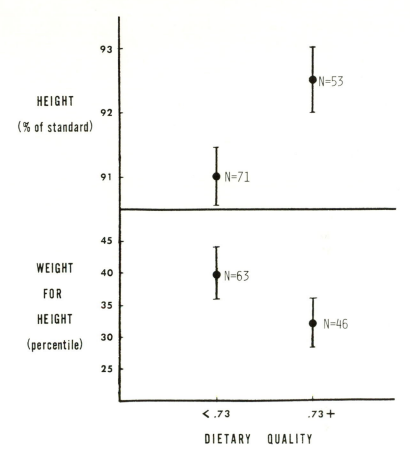

Figure 7.   Relationship of height and weight-for-height to dietary quality (means
+ / − one standard error). Student *t* test of difference between means
significant for height (p < 0.05), but not for weight-for-height
(p < 0.10). (Data for children of *socios* and from Teco are combined for
this analysis).

A possible explanation for this is that weight-for-height—a measure
of fatness—is a more labile variable, and may thus be more strongly
influenced by non-dietary factors such as illness. Diarrhea, for
example, can lower a child's relative weight-for-height in a matter of
days.

A summary of these results is given in Figure 8, which shows all of
the relationships found to be significant for socios (solid lines) and
families in Teco (dashed lines). It appears that the chain of factors on

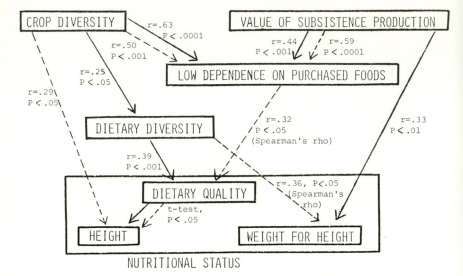

Figure 8.   Demonstrated relations between crop diversity, dependence on purchased foods, and nutritional status. Solid lines = *socios*, dashed lines = Teco. r = correlation coefficient. Although group comparisons are more appropriate for dietary recall data than are correlations (see footnote 3), correlation coefficients are listed here in order to provide a means by which to compare the relative strength of the associations found.

the left, from crop diversity to dietary diversity to dietary quality, is best illustrated by socio families, while the chain of factors on the right, relating the degree of dependence on purchased foods to dietary quality, is best illustrated by families in Teco.[6] Dependence on purchased foods may have more of a negative effect in Teco for two reasons. First, because Teco is more isolated than the poblados of the Plan, the variety of foods available in the local stores is extremely limited; and second, since work is not available on a regular basis, wages generally fluctuate more from week to week for families in Teco than they do for socios, which may make dependence on cash for purchasing food more hazardous.

These results are based on a statistical analysis, and their interpretation is, therefore, subject to the limitations of such an approach. Correlations and associations do not necessarily demonstrate causation. Nevertheless, when taken together, they do imply that families in this study who have maintained a greater degree of self-sufficiency are better off nutritionally than those who have not.

Less reliance on purchased foods may be especially advantageous in an area such as Tabasco, where prices for food are so high relative to wages. The relationships found between crop diversity, dietary diversity, and dietary quality for socio families point to the importance of growing a variety of crops, unless foods of equivalent value are both physically and economically accessible to rural families. These results thus lend a nutritional argument to the other advantages (ecological and economic) of maintaining a diversified agricultural system.

A similar result was found by Marchione (1977) in a study of child nutrition in Jamaica. When market conditions forced small producers to return to greater subsistence farming, nutritional status of children improved. The complex interactions between national food policy, market variables and sociocultural factors were considered responsible for this result.

AN EXAMPLE: SUGAR CONSUMPTION

A good illustration of the kind of dietary change that may accompany agricultural change is the trend toward increased sugar consumption. A native Tabascan drink, pozol, is highly nutritious, containing substantial amounts of protein and other nutrients. It is even more

TABLE V

Nutritional content of *pozol* versus soft drinks[a]

	Pozol	Soft drinks
Energy (kcal)	264	120
Protein (gm)	6.3	0
Calcium (mg)	117	0
Iron (mg)	2.3	0
Thiamine (mg)	0.23	0
Riboflavin (mg)	0.07	0
Niacin (mg)	1.0	0
Ascorbic acid (mg)	0	0
Vitamin A (μg Eq)	0	0

[a] Per 250 ml serving. Nutrient composition calculated from Hernandez, Chavez and Bourges 1977; composition of *pozol* based on 125 g corn *masa* and 5 g cacao per 250 ml serving.

valuable when allowed to ferment into *pozol agrio*, a common practice in the area, since the microbes that cause fermentation increase its protein, niacin and riboflavin content and also improve protein quality (Ulloa 1974). In recent years, as families have produced less and less corn and cacao for home consumption, pozol has been increasingly replaced by soft drinks which are sold in every store in every village of the area.[7] Table V compares the nutritional content of a 250 ml (approx 8.5 oz.) serving of pozol with the same quantity of a soft drink. In an area such as Tabasco where the weather is so hot, liquids can make up a substantial portion of a child's daily dietary intake; a child may drink as much as a quart of pozol in one day. Clearly, if soft drinks are substituted for pozol, the quality of the child's diet will decline.

Figure 9 shows the average consumption of soft drinks, table sugar and total sugar for children of socios, libres and families in Teco. Soft drink consumption is higher in the Plan, while use of table sugar is higher in Teco. When the two are combined, there are no significant differences in total sugar consumption in the three groups.

The negative nutritional impact of the trend toward increased sugar consumption is illustrated by Figure 10. For children in this study, there is a significant negative correlation between sugar consumption and height, even when other variables that might influence height are controlled for.[8] It is likely that sugary foods such as soft drinks are replacing foods with superior nutritional value in the diet. This hypothesis is supported by the negative correlation between sugar consumption and protein intake ($r = 0.16$; $p < 0.05$). This effect on dietary adequacy could then have an impact on the more long-term measures of nutritional status, such as height.

Sugar and soft drinks are just one example of the poor nutritional quality of many of the foods that are commonly purchased by families in the study. Another trend is the replacement of tortillas with sweet rolls, white bread, crackers and cookies. Such inexpensive foods are the only items that most of the families can afford; the nutritionally superior foods such as fruits and vegetables, eggs, milk, chicken, fish and meat which used to be produced by most families ten or twenty years ago are now scarce or extremely expensive or both. Beef, for example, is usually sold only once or perhaps twice a week in the villages, and even then there is not enough for everyone. One socio woman told me that "Para comprar un kilo de carne, tiene que perder una noche" (In order to buy a kilo of meat you have to lose a night's sleep), because meat is sold at 1 a.m. in the poblado and is all gone by

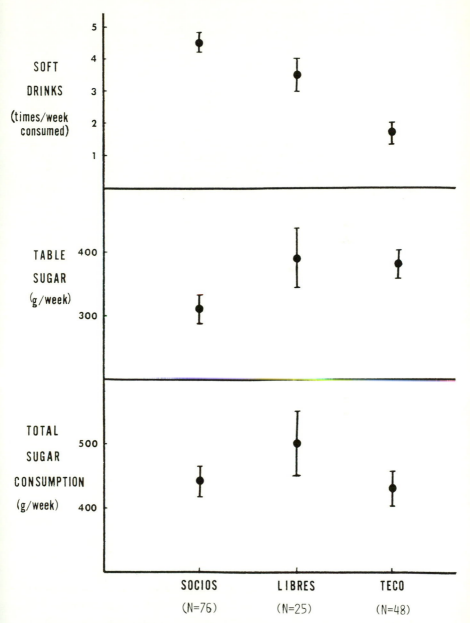

Figure 9. Soft drink, table sugar, and total sugar (sucrose) consumption of families (means +/− one standard error) in Tabasco, Mexico.

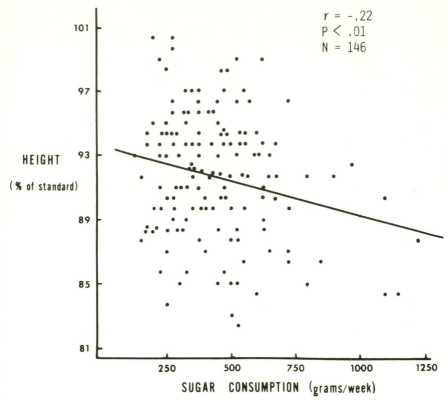

Figure 10.   Relationship of height (expressed as % of Mexican standard) to total
sugar consumption in study children. r = correlation coefficient.

4 a.m. This is particularly ironic in view of the fact that 65% of the
land in the Plan Chontalpa is devoted to cattle production. Nearly all
of the beef produced is exported out of Tabasco to central Mexico.

INCOME, SUBSISTENCE PRODUCTION AND NUTRITIONAL
STATUS

It is sometimes argued that the shift from subsistence to commercial
production should make little difference to diet and nutrition if
income levels are raised in the process. This is obviously true in cases
where the increased income more than compensates for the loss of
home-grown food. But for families in this study the relationships

TABLE VI

Correlations between indicators of nutritional status and per capita income in study families

	SOCIOS	LIBRES	TECO
Weight	− 0.01	0.64[a]	− 0.44[a]
Height	0.22	0.48[b]	− 0.27
Weight for height	− 0.17	0.50[b]	− 0.37[b]
Head circumference	0.02	0.29	− 0.46[a]
Arm circumference	0.03	0.72[a]	− 0.21
Triceps skinfold	− 0.05	0.17	− 0.05
Dietary quality	0.18	0.64[a]	0.13
Energy intake	0.10	0.66[a]	0.13
Protein intake	0.08	0.75[a]	0.04
Biochemical measures	− 0.02	0.26	− 0.24

[a] $p < 0.01$.
[b] $p < 0.05$.

between income, subsistence production and nutrition are more complicated. For libre families, income and nutritional status are clearly related. Because libres have little or no land available to them for the production of food, their diet and nutritional status depend primarily on cash income. Table VI shows that there is a significant positive correlation between most of the measures of nutritional status and *per capita* income for libre children. However, this is not the case for socio children and children in Teco. There are no significant correlations between nutritional status and income for socios, while in Teco there seems to be a negative relationship, if anything. The lack of a positive relationship between income and nutritional status for these groups emphasizes the inadequacy of using income measures alone in a situation in which subsistence production plays an important role in determining food consumption. Although the value of subsistence production has been included in the measure of *per capita* income used in Table VI, it is likely that the cash value of crops grown for home consumption does not fully represent their nutritional value. In other words, the contribution of subsistence crops to the diet may be greater than just their monetary value. Furthermore, if crops grown for home consumption were to be purchased instead, the cash required to buy them would be greater than their monetary value if sold, due to the markup in price between what the farmer receives and what the consumer pays.

In most nutrition surveys in which a positive relationship between economic status and nutrition has been found, there is a wider range of economic status among the families surveyed than in this study. It is not surprising that increased malnutrition is found in groups of landless families when compared with families who do have land, for example. In such broad-scale studies, families can be separated into relatively distinct categories of economic status, and it is usually found that economic status is the most important influence on diet and nutrition. The libre families in this study are an example of such a case. There are basically two classes of libres: families in which the father is a skilled or semi-skilled laborer with regular employment (truck or bus drivers, mechanics, carpenters and construction workers), and families in which the father is an agricultural day laborer with variable employment and low wages. The variance in income is thus very great among libre families. However, when the families in a study are seemingly more homogeneous in economic status, such as the socios and families in Teco in this study, it is less probable that a relationship between income and nutritional status can be demonstrated, especially if the situation is complicated by the difficulty of assessing the contribution of subsistence production.

It is quite possible that there are too few families in this study with incomes high enough to make a real difference to diet and nutrition. It has been suggested that when families undergo the change from subsistence to cash cropping, their resulting income is still marginal with respect to their needs for cash, meaning diet and nutrition would be expected to decline. Only when families reach a higher level of economic status would their diets improve sufficiently to show the positive effect of increased income. It is, therefore, in the transitional period that one would find the greatest degree of malnutrition. This kind of categorization may apply to the families in this study. When children of socio families whose *per capita* incomes are higher than 250 pesos per week (the top 25%) are compared with those whose incomes are less than 250 pesos per week, children in the top income bracket are significantly taller (94.4 cm versus 91.3 cm, Mann-Whitney U test, $p < 0.01$). The same result is not found for the measures of fatness, perhaps because fatness is a more variable characteristic as explained in a previous section. If the same income categories are used for children in Teco, there are only four families with *per capita* incomes greater than 250 pesos per week, too few for any valid comparisons to be made. This indicates that there is probably not enough variation in economic status within Teco to demonstrate any positive relationship of income level and nutrition.

There are a number of other reasons why families in Teco with somewhat higher incomes would not necessarily be nutritionally better off. First, a higher income would improve diets only if the extra income is spent on food. Such is the case for socio families. There is a significant correlation between *per capita* income and the amount of money spent for food (r = 0.31, p < 0.01). There is no such relationship for families in Teco, however (r = 0.14, p > 0.10). Second, even if cash income is spent for food, purchased foods may be of poor nutritional value, as illustrated by the example of sugar consumption described previously. In this study there is a positive correlation between income and the frequency of consumption of soft drinks (r = 0.56, p < 0.01) since families with extra cash are likely to spend it on such items rather than on other foods which are considerably more expensive. Fruits and vegetables, for example, are often not consumed at all unless they are grown in the family garden, partly because they are not readily available but also because they cost more than even the wealthier families are willing to spend. Children of families who produce some or all of their own fruit in the solar consume fruit more often than children of families who buy all their fruit (3.4 vs. 2.4 times per week, Student t test, p < 0.05). These reults indicate that merely replacing food with cash does not guarantee that dietary quality will remain the same.

To summarize, cash income is not always a reliable predictor of nutritional status in situations in which subsistence production continues to make a substantial contribution to the diet and in which very few families have incomes that enable them to purchase an adequate diet.

THE SHIFT TO CATTLE PRODUCTION IN TECOMINOACAN

Although the families of Teco have long been at least partially involved in the market economy, in recent years there has been a marked increase in the amount of land devoted to cattle pasture. As explained previously, nearly all of the fertile alluvial soils which were the original lands of Teco were expropriated by the Plan Chontalpa, and the land that remains is the poorer quality savanna-type soils. Many villagers complain that the savanna soils are only good for pasture grasses: "Pasto es lo maś que se da." (Pasture is the most that can grow there.) They say that it dries out too much during the dry season and that more often than not they lose nearly all of their crops. Others claim that the most serious problem is the invasion of croplands by seeds or vegetative runners of the weedy grasses from neighboring cattle pastures. Any crops that they attempt to grow cannot compete with the grasses, and are choked out almost immediately. Many families mentioned that ten or so years ago the yields of corn, beans and other crops were substantial—almost as high as yields from the alluvial soils—but now they can no longer produce much of anything due to the *zacate* (weedy grass). As a result, more and more families are abandoning the cultivation of basic crops and converting their land to cattle pasture. In many cases they have no other alternative, as there seems to be a chain reaction in the conversion to pasture. A neighbor decides to switch to pasture, which reduces the yields of the next neighbor's crops, who then decides to switch to pasture as well, and so on.

Even though many families in Teco now have land in pasture, it has been difficult for them to make use of their land for cattle production. While 24 families in the study have some pasture land, only 16 have any cattle and only half of those have more than ten cattle. This is partly due to the difficulty of obtaining financial credit to purchase cattle initially. One group of villagers was able to get credit from a bank to start up production in 1971, but aside from that instance most families have been unable to obtain any loans. Without credit, the costs of cattle production are much too high for most villagers to afford. Just the fencing material necessary to enclose a pasture of ten hectares (25 acres) costs approximately 11,000 pesos —almost as much as an entire year's income for many families. Families who do own cattle often do not have adequate fencing or are unable to maintain it properly, and neighbors complain that their crops are trampled by wandering cattle. If pastures are planted in the higher-quality artificial (not native) grasses, the cost of seed for a ten-hectare pasture may run to 2400 pesos. Because nearly all of the work required to create and maintain the pasture is done using manual labor, a family usually has to hire outside laborers. To clear a field prior to planting with pasture grass, for example, may require 8–16 days of labor per hectare (3–6.5 days per acre), depending on the height of the second-growth vegetation. The cost of labor to accomplish this task may be as much as 2500 pesos per hectare. Once the pasture is planted, it must be kept free of non-edible vegetation, which requires about six days of labor per hectare (2.5 days per acre) each year. Other costs include expenditures for horses, herbicides, veterinary medicines, fence poles and other equipment. Establishing a sizable herd (perhaps 20–30 head) may, therefore, take ten or more years.

One of the objectives of this study was to determine what effect, if any, this shift from a more subsistence-based agriculture to cattle pasture has had on diet and nutrition in Teco. Because longitudinal information is not available, the only way to investigate this question is to compare families who have cattle pasture with those who do not. Table VII lists the measures of anthropometry and dietary quality for children of families with and without pasture (excluding non-agricultural families), along with average family income, *per capita* income, land availability and value of subsistence production in the

TABLE VII

Nutritional status of children in Tecominoacán: families with and without pasture[a]

	Families with Pasture	Families without Pasture
Weight	81.3	84.2
(% of standard)	(20)	(12)
Height	91.6	90.0
(% of standard)	(24)	(16)
Weight for height	25.5	42.2
(percentile)	(20)	(12)
Head circumference	96.0	96.1
(% of standard)	(20)	(12)
Arm circumference	97.7	99.8
(% of standard)	(20)	(12)
Triceps skinfold	103.9	102.4
(% of standard)	(20)	(12)
Dietary quality	82.6	84.4
(mean % of RDA)	(24)	(16)
Family income	859	537[b]
(pesos/week)	(23)	(14)
Percapita income	134	111
(pesos/week)	(23)	(14)
Land availability	22.8	5.5[c]
(hectares)	(24)	(16)
Value of subsistence	6813	3757
production (pesos)	(23)	(15)

[a] Values of N in parentheses.
[b] t test of difference between means significant, $p < 0.05$.
[c] t test of difference between means significant, $p < 0.001$.

F

two groups. Most of the families with pasture also produce crops on their own or rented land, which accounts for their high mean value of subsistence production. For this reason, a comparison cannot be made between subsistence production vs. cattle production, as both groups still grow some subsistence crops. Rather, the question must be posed as to whether families who have cattle pasture in addition to other crops are any better (or worse) off than families without pasture. Table VII shows that there are no significant differences in any of the measures of nutritional status of children in the two groups. Families with pasture, however, have significantly more land and higher family incomes than families without pasture. The same results are found if the comparison is made between families with cattle and families without cattle rather than with and without pasture. It, therefore, appears that even though families switching to cattle production have more land and more cash than other families, there is no difference in the diet and nutrition of their children.

As mentioned, the shift toward increased cattle production in Teco cannot be viewed as a clearcut transformation from subsistence to commercial agriculture, as some families with cattle continue to grow subsistence crops, but the comparison can be used to evaluate the common assumption that a rise in income accompanying the adoption of commercial production will automatically lead to improved nutrition. In this particular example, it has not. Furthermore, the expansion of cattle pasture in Teco has had repercussions that have affected the entire village, not just those families who have cattle. Even families who do not wish to convert their land to pasture have difficulty raising crops due to the invasion of pasture grasses. The villagers of Tecominoacán are trapped in a difficult situation. Cattle production is economically unfeasible for them, but because they can no longer produce their own food, they must rely on irregular sources of outside wage labor, even though they may have 20 hectares (50 acres) or more of land.

SUBSISTENCE PRODUCTION, CASH CROPPING, AND WAGE LABOR

The results of the cross-sectional comparisons discussed in the previous sections confirm the historical trends described to us by the families we interviewed. Figure 11 illustrates the percentage of families who felt that there was more, less or the same amount to eat

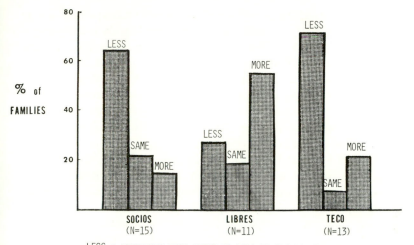

Figure 11.   Changes in food consumption as perceived by study families over the last ten years.

now as compared with ten years ago (the data are from the subsample of 44 families). A majority of socios and families in Teco felt that the situation had deteriorated, especially for fruits and vegetables, meat, wild game and fish. (Only for the libre families did a majority feel that there was more to eat now. Most libre families were landless laborers ten years ago, so this result is not surprising). In general, most families felt that the decline in their diet was due to the reduction in the production of subsistence crops and in the availability of wild sources of food. Many of them make comments such as "Antes no sufríamos—era mejor producir lo que necesitaba—siempre había comida. Ya no se puede—solo cuando hay dinero hay comida." (Before we didn't suffer—it was better to produce what we needed —there was always food to eat. Now it is no longer possible —only when there is money is there food to eat.) The ability to be self-sufficient in food is seen as a real advantage; some would rather grow subsistence crops than cash crops because they feel that "Dinero no me dura." (Money doesn't last me.) Yet despite the fact that they recognize the value of producing food for home consumption, many families are switching to cash crops or cattle production. If this change is not beneficial to them, why is there a trend toward commercial production?

In Teco the conversion to cattle pasture can be understood as being in large part due to environmental causes—the loss of their alluvial soils plus the rapid invasion and establishment of pasture grasses on the land that remains. For socios of the Plan Chontalpa, the issue is more complex: Why are more and more socios devoting their 2-hectare (5 acre) plot to cash crops? The reasons for this trend have to do with the structure of agricultural production in the Plan (the difficulties of maintaining a dual livelihood of individual production and wage labor on the collective land).

Socios of the Plan are faced with a dilemma. If they work for wages on the collective land, they are usually left with too little time or energy to successfully raise food for their family on their individual plot. Although there are sometimes weeks in which no collective work is available, such periods are not predictable far enough in advance to permit a socio to devote that time to the family plot.[8] Because working on the collective land is required to retain membership in the ejidos, the family plot is given second priority; but the wages earned from working within the ejidos are not sufficient to feed a family adequately in many cases. Socios must, therefore, earn more cash in some way, either by working outside of the ejidos or by producing cash crops. Although working outside the Plan can result in a socio family losing its ejidal rights, some families have chosen this route. One ex-socio who now drives a truck told me, "No trabajaba en el colectivo porque no podía mantener la familia. Me quitaron los derechos y la parcela." (I didn't work in the collective because I couldn't support my family. They took away ejidal rights and the family plot.) Work outside the Plan can be very disruptive to the family, however, due to the enormous time commitment that is usually required. If a socio works for a company in the nearest town (Cárdenas), for example, his daily schedule might be as follows: Up at 3 a.m., arrive at work (by bus) at 5 a.m., begin work at 6 a.m., work until 5 p.m., return home at 7 p.m. A number of socios have begun to travel even farther to find construction work or other jobs; in these cases they leave on Sunday afternoon and work through until Saturday morning, returning home for only one day each week.

The other alternative for acquiring more cash, growing a cash crop, is more commonly chosen. The time constraints imposed by wage labor, plus the desire to be more involved in modern agriculture, have led many socios to grow cacao, bananas or sugar cane on their family plots instead of subsistence crops. As I will explain below, the production of these crops can usually be more easily incorporated into

TABLE VIII

Estimated labor required to plant two hectares (five acres) of corn[a] in a Mexican community

Procedure	8-hour days
Clearing plot for planting, using machete	16–40
Planting, using dibble stick	6–12
Weeding, using machete	12–16
"Doubling" (bending stalks over to dry) and weeding	6–20
Harvest (depending on yield)	8–20
Total for one growing season	48–108
If planted during both growing seasons	96–216

[a] Based on detailed interviews with 11 families.

a combination of wage labor and work on the family plot than the production of subsistence crops using manual labor. The expenses and risks of cash crops are much greater, however.

In Table VIII the number of days (8 hours) of labor required to plant 2 hectares of just corn are listed. As the figures illustrate, the time investment required to produce corn on the family plot is substantial. Furthermore, growing corn successfully depends to a great extent on the timing of planting, given the variable weather (particularly rainfall) in the area. Clearing the plot, which can take more than a month of labor, must be completed prior to planting, but not too far in advance or the weeds will once again cover the plot. This necessity for scheduling creates a time constraint that makes it difficult for a socio to work on the collective land 4–6 days per week and still produce subsistence crops. Farmers in the area used to rely on reciprocal forms of group labor for tasks such as clearing and planting; but as I will discuss later, such forms of cooperative labor have virtually disappeared with the expansion of commercial agriculture.

One way out of this time constraint is for the socio to hire someone to plow the family plot rather than clear the vegetation by hand. This is only feasible, however, for families whose plots are close enough to the road for heavy equipment to be used. Even then, the cost of having the plot plowed is quite high (2000–6000 pesos for two hectares [five acres]), and families may have to wait several weeks for their plot to be prepared, which may delay planting and cause the crop to be jeopardized. If the corn produced is to be used for home consumption, the cost of mechanization must be paid out of the socio's already inadequate wages; but the alternative of selling the corn is not very lucrative. Given that the price for corn averaged about 3 pesos/kg in 1978, a typical yield of 2800 kg for 2 hectares (6200 lbs for 5 acres) would be worth about 8400 pesos. Since most of the socios recognize that the other cash crops can earn much more than this, producing corn as a cash crop is not very common.

The other cash crops generally require a substantial amount of labor, but the scheduling of tasks is usually not as critical as for the basic subsistence crops. Cacao production requires approximately 100–140 days of labor per year for a two hectare plot (for weeding, fertilizing, fumigation, pruning and harvesting). While this is not much less than the labor needed for corn production, most of those who grow cacao have only one hectare (2.5 acres) of mature trees, which would require only 50–70 days of labor per year. Unlike the labor needed for corn production, most of the work for cacao production can be spread out over the entire year, and timing is not as critical.

Production of bananas as a cash crop may require a large labor input if weeding is done by hand with a machete. Those who produce bananas usually use an herbicide, however, which decreases greatly the labor required. Once the banana starts have been planted, harvesting can occur at any time throughout the year, allowing a socio to spend as little or as much time as is available on the family plot. Growing bananas as a cash crop is only possible for families with plots close to a road so that trucks can pick up the harvested fruit.

Sugarcane produced by individual socios is grown under contract with the local sugarmill. While sugarcane production requires a great deal of labor (about 150 days per year for weeding, fertilizing and harvesting a two-hectare [five-acre] plot) and timing is quite crucial, many of the socios who grow sugarcane use the credit given by the sugarmill to pay other laborers to work their family plot. As a result, the socio is in a position to continue working on the collective land as well as producing a cash crop.

Whether a family chooses to produce a cash crop or to use mechanization for corn production, the expenses, and therefore the risks, can be quite high. All of the cash crops require the use of fertilizers, insecticides and/or herbicides, which are expensive and necessitate the purchase of a back-carried sprayer costing 1500–2000 pesos. Despite the use of insecticides, diseases or pests commonly afflict a portion of the crop. If the weather has been rainy and the fields are too muddy for the sugarcane trucks to enter, the cut cane ferments and is no longer acceptable to the sugarmill. As a result of these difficulties, only families who can afford to take on the added risks are able to go into cash cropping. Ironically, even the production of subsistence crops is easier for wealthier families, as they can afford to hire outside labor to work their plot or to have it plowed. The data collected on wages (not including crop income) of socios in this study show that those who produce cash crops or pay to have their land cleared earn more than those who do not. The difficulty of the dual livelihood of wage labor plus individual production is, therefore, felt most by the poorer families who neither earn enough from wage labor nor have the capital necessary to produce a cash crop or to use machinery for the production of subsistence crops.

The transformation from growing subsistence crops to commercial production on the family level is thus intimately tied to the transformation from peasant farmers to wage laborers that has been fostered by the Plan Chontalpa. Of course, it is likely that even without the Plan, farmers in the area would have increased their production of cash crops. But the Plan has effectively undermined the ability of the farmers to continue producing food for home consumption (see Dewey 1980b), substituting instead the meager wages of work on collective production which for the majority of socios provide neither an adequate diet nor sufficient income to successfully produce a cash crop.

While the focus here has been on the impact of this transformation on diet and nutrition, it is important to point out that the effects are more far-reaching, changing the nature of social relations among peasant families as well as their health status. Both in Teco and in the Plan, the production of cattle or cash crops often requires the hiring of laborers, a practice previously limited to much larger scale farmers in the area. In the past, mutual exchange of labor on others' plots was the universal way in which very time-consuming tasks were accomplished. Gudeman (1978) describes the same patterns among small farmers in Panama and the loss of reciprocal labor that resulted from the shift toward sugarcane production in the community he studied. A greater involvement in the outside market caused a transition from the "household-household integration" promoted by cooperative labor forms to a "household-market integration" through the hiring of labor. The same situation exists in Tabasco, in which cash has come to replace cooperation and inter-household ties are weakened with the increased demands for wage labor.

As in other cases where cash cropping has diminished the status of peasant women (Stavrakis and Marshall 1978), the changes in agricultural production in the Chontalpa area have affected the sexual division of labor. Because most families prior to the Plan Chontalpa lived on their family plots rather than in a centralized village, work on the crops could be shared by all of the members of the family. Now, however, the family plots in the Plan Chontalpa are usually too far from the houses for women with small children to be able to contribute much labor. In addition, commercial production, which involves decisions about spending and earning cash, is typically within the domain of the men, thereby reducing the economic power of the woman within the household. This tendency is further exacerbated if the man is gone for long hours each day or for an entire week in order to earn money. As a result, many of the women we interviewed said that they very rarely left the house and consequently felt extremely isolated from the outside world.

The loss of reciprocal labor and the changed status of women are just two examples of the kinds of changes effected by the conversion from subsistence to commercial agriculture. I have included them here to point out that social consequences, as well as nutritional and health-related consequences, must be considered in any assessment of the impact of agricultural change.

CONCLUSION

This article has focused on some of the effects of the transformation from subsistence to commercial agriculture in the Chontalpa area. Because of the complexity of the social and economic relations both in the Plan Chontalpa and in Teco, families cannot be separated into easily distinguishable groups of subsistence and commercial farmers, making a simple comparative test to determine how increased cash cropping influences diet and nutrition impossible. Rather, I have attempted to examine the effects of some of the changes that generally accompany the shift to commercial agriculture, namely, the reduction in the degree of self-sufficiency in food among peasant families. The results of this study indicate that when families are compared with respect to the crop diversity of their family plots and their degree of dependence on purchased foods, children of families who are more self-sufficient are better off nutritionally. As mentioned, this conclusion must be interpreted within the context of the particular situation that exists in the Chontalpa area, which has a combination of high food prices and low agricultural wages and structural constraints imposed by the Plan Chontalpa. While these results point out the nutritional advantages of self-sufficiency in food for peasant families in Tabasco, it is certainly not my intent to advocate a massive return to subsistence farming. The issues are far more complex than simply whether or not a family produces or buys its food. The central issue is how and under what conditions agricultural change occurs. What is needed is a philosophy of development that is likely to lead to a more progressive form of agricultural change than that which has occurred in Tabasco.

NOTES

1. The collaborators whose contributions made the study possible were: interviews in the Plan Chontalpa, Regina Gonzalez García; and interviews in Tecominoacán. Janifer Burns, Kathleen Foote Durham, William Durham; medical assessment, Agustín López Hernandez, Martha Alicia Dias de F. Acknowledgment is made to J. Vandermeer, R. Ford, M. Taussig, G. Owen, B. Hazlett and W. Durham for their helpful comments. Financial support was provided by a Social Science Research Council International Doctoral Research Fellowship, a National Science Foundation Doctoral Dissertation Grant (#BNS 78-05517), Rackham Block Grants from the University of Michigan and by a Grant-in-Aid of Research from Sigma Xi, the Scientific Research Society of North America. The research was carried out in collaboration with the Colegio Superior de Agricultura Tropical in Cárdenas, Tabasco, and with the permission of the Secretaría de Recursos Hidráulicos.

2. Children younger than 2 years were not included because of the difficulty of measuring dietary intake prior to weaning. Information on infant feeding practices was collected, however, and is described elsewhere (Dewey 1980a).

3. Although many investigators have pointed out the inadequacies of the 24-hour recall (e.g. Garn et al. 1978), it is the only feasible method for assessing dietary intake in a survey of this size. Families in the study would not generally have been able to keep dietary records, either because of illiteracy or lack of motivation, and careful weighing of all foods consumed would have been extremely time-consuming. By asking the mother of each child to recall the foods consumed the previous day, and then returning the next day for a second recall, a better picture of dietary intake was obtained than with a single recall. Care was taken to probe for foods and drinks that the mother may have forgotten to mention. Nevertheless, there are many sources of error in any estimate of dietary intake (e.g., the memory of the mother, intentional or unintentional bias in reporting amounts consumed, interviewer differences, etc.), in addition to the normal day-to-day variation in food consumption. Because of this daily variation, it is more appropriate to compare dietary intake data of *groups* of children rather than *individuals*. For this reason, all of the tables and figures presented in this paper which include dietary recall measures are based on group comparisons.

4. Statistical analysis of the data was performed using primarily parametric statistics such as Student t, analysis of variance and regression analysis. The assumptions of normality and homogeneity of variances were checked for each analysis, and

whenever the assumptions were not met, non-parametric statistics such as Mann-Whitney U and Kruskal-Wallis were utilized instead.

5. Stepwise multiple regression analyses were performed for each dependent variable in order to control for the influence of all potenially confounding independent variables. The information is summarized below for all independent variables with a partial correlation coefficient significant at the $P \le 0.05$ level. The summaries correspond to the relationships shown in Figures 5–7. Figures 5–7, however, are based on group comparisons, which are more appropriate when making statistical inferences based on dietary recall data.

Figure 5: *Dietary diversity*: regression with crop diversity of plot, crop diversity of *solar*, dependence on purchased foods, value of subsistence production, family income, per capita income, maternal educational level:

	Variables included	Partial correlation coefficient	Significance
Socios	Crop diversity of plot	0.23	P = 0.05
Teco	none		

Figure 6: *Dietary quality*: regression with above variables pus dietary diversity, presence of clinic, and sex of child (after adjusting for normal sex-related differences by using appropriate standards):

	Variables included	Partial correlation coefficient	Significance
Socios	Dietary diversity	0.46	P < 0.001
	Per capita income	0.34	P < 0.01
	Sex (girls' diets poorer)	0.28	P < 0.05
	Clinic (diets better in poblados with clinic)	0.26	P < 0.05
Teco	Dependence on purchased foods	− 0.40	P < 0.01

Figure 7: *Weight for height*: regression with maternal weight, maternal height, maternal educational level, dietary diversity, dietary quality, caloric intake, crop diversity, value of subsistence production, family income, per capita income, and sex of child:

	Variables included	Partial correlation coefficient	Significance
Total sample	Sex (girls thinner)	0.29	P < 0.01
	Family income	− 0.28	P < 0.01
	Maternal weight	0.26	P < 0.01

Height: regression with maternal weight, maternal height, dietary diversity, dietary quality, caloric intake, crop diversity, family income, per capita income, sex of child, and total sugar consumption:

	Variables included	Partial correlation coefficient	Significance
Total sample	Family income	0.26	P < 0.01
	Maternal height	0.22	P < 0.05
	Total sugar consumption	− 0.21	P < 0.05

Other potentially confounding variables (family size, birth order of the child) were not found to be related to diet or nutritional status in this study.

6. It should be noted that not all of the significant relationships follow the pathways indicated in Figure 8. Height of children in Teco is directly related to crop diversity, but not to dietary diversity, while weight-for-height of *socio* children is related to the value of subsistence production, but not to dietary quality. The situation is therefore more complicated than shown in Figure 3.

7. When asked about trends in consumption of selected foods over the past ten years, nearly all of the families interviewed (the subsample of 44) stated that their consumption of pozol had decreased while soft drink consumption had increased.

8. It is legitimate to use a correlation rather than a group comparison here because data on sugar consumption are based on food frequency information rather than dietary recalls, thus eliminating the daily variation source of error.

REFERENCES

Beaudry-Darisme, Micheline N.; Lesly C. Hayes-Blend and A. G. Van Veen. 1972. The Application of Sociological Research Methods to Food and Nutrition Problems on a Caribbean Island. *Ecology of Food and Nutrition* 1:103–119.

Dewey, Kathryn G. 1979. Agricultural Development, Diet and Nutrition. *Ecology of Food and Nutrition* 8:265–273.

——— 1980a. *The Ecology of Agricultural Change and Child Nutrition: A Case Study in Southern Mexico*. Ph. D. dissertation, University of Michigan.

——— 1980b. The Impact of Agricultural Development on Child Nutrition in Tabasco, Mexico. *Medical Anthropology* 4(1):21–54.

Fleuret, Patrick and Ann Fleuret. 1980. Nutrition, Consumption, and Agricultural Change. *Human Organization* 39(3):250–260.

Garn, Stanley M., Frances A. Larkin and Patricia E. Cole. 1978. The Real Problem with 1-day Diet Records. *American Journal of Clinical Nutrition* 31(7):1114–1116.

Gross, Daniel R. and Barbara A. Underwood. 1971. Technological Change and Caloric Costs: Sisal Agriculture in Brazil. *American Anthropologist* 73:725–740.

Gudeman, Stephen. 1978. *The Demise of a Rural Economy: From Subsistence to Capitalism in a Latin American Village*. London: Routledge & Kegan Paul

Hernandez, M., C. P. Hidalgo, J. R. Hernandez, H. Madrigal and A. Chavez. 1974. Effect of Economic Growth on Nutrition in a Tropical Community. *Ecology of Food and Nutrition* 3:283–291.

Hernandez, M., A. Chavez and H. Bourges. 1977. *Valor nutritivo de los alimentos mexicanos*. México: Instituto Nacional de la Nutrición

Krebs, Charles J. 1972. *Ecology*. New York: Harper & Row

Marchione, Thomas J. 1977. Food and Nutrition in Self-Reliant National Development: The Impact on Child Nutrition of Jamaican Government Policy. *Medical Anthropology* 1:57–79.

Palmer, Ingrid. 1974. The Social and Economic Implications of Large-Scale Introduction of New Varieties of Foodgrain: Summary of Conclusions of a Global Project. (United Nations Research Institute for Social Development).

Pearse, Andrew C. 1975. *The Latin American Peasant*. London: Frank Cass

Ramos Galvan, R. (ed.). 1975. Somatometría pediátrica estudio semilongitudinal en niños de la Ciudad de México. *Archivos de Investigación Médica* 6 (Suppl. 1).

Stavenhagen, Rodolfo. 1978. Capitalism and the Peasantry in Mexico. *Latin American Perspectives* 8(V)(3):27–37

Stavrakis, O. and M. L. Marshall. 1978. Women, Agriculture and Development in the

Maya Lowlands: Profit or Progress. Paper presented at a conference on Women and Food, January 1978, Tucson, Arizona.

Taussig. Michael 1978. Peasant Economies and the Development of Capitalist Agriculture in the Cauca Valley, Colombia. *Latin American Perspectives* 18(V)(3):62–90

Teitelbaum, Joel. 1977. Human Versus Animal Nutrition: A "Development" Project among Fulani Cattlekeepers of the Sahel of Senegal. In *Nutrition and Anthropology in Action*, ed. by T. Fitzgerald. Assen: Van Gorcum

Ulloa, M. 1974. Mycofloral Succession in *Pozol* from Tabasco, Mexico. *Boletin. Sociedad Mexicana de Micología* 8:17–48.

Wasserstrom, Robert. 1978. Population Growth and Economic Development in Chiapas, 1524–1975. *Human Ecology* 6(2):127–143.

PART III
ADAPTATION TO ECOSYSTEMS

CHAPTER 7

THE BANANA IN USAMBARA[1]

ANNE FLEURET AND PATRICK FLEURET
The American University
Washington, D.C.
Africa Bureau
U.S.A.I.D.
Washington, D.C.

ABSTRACT

The dietary, nutritional and cultural role of the banana (*Musa* sp.) among the Shambaa people of the Usambara Mountains, north-eastern Tanzania, is explored. Using archival, historical and ethnographic data, the authors show that in pre-colonial times the banana was paramount in Shambaa diet and as a source of nutrients, and that agricultural practices were specifically designed to foster banana production. Although other cultivated foodstuffs have now relegated the banana to a secondary place in contemporary Shambaa diet and nutrition, it remains important in trade and maketing and for various nonfood uses, which are also discussed.

INTRODUCTION

The banana is one of the less common staple foods and its use as the primary source of calories in the diet is limited mainly to areas of tropical Africa. Bananas cannot be cultivated in temperate zones, and the fruits that are seen in Western countries are rather standardized products grown solely for the export market. Commercial production is concentrated in Caribbean and Latin American countries and has depended on a very few clones that produce a fruit with a high sugar content which is consumed in the West as a dessert or snack item. Yet patterns of consumption of the banana in countries like the United States are certainly not typical of the world as a whole. Of the

145

estimated 20 million tons of the fruit produced annually worldwide, only 15% enters the export trade (Simmonds 1976:211). The remaining 17 million tons are consumed locally in the areas of production. In some tropical countries the banana is a staple article of the diet and is in some areas (such as Uganda) a "cultural super-food" (Jelliffe 1966)—the source of the bulk of the calories and other nutrients consumed by the population. We intend here to provide an historical and contemporary account of banana use in one area of East Africa—the Usambara Mountains of Tanzania—where bananas have been culturally, economically and nutritionally significant to the local people for a very long time.

THE SETTING

The West Usambara Mountains are largely contained within the boundaries of Lushoto District, Tanga Region, northeastern Tanzania. The mountains rise abruptly from the surrounding plains; a sheer escarpment some 500 meters (m) (1650 ft.) high rings the highland area. In the interior portions of the highlands individual peaks rise as high as 2000 m (6500 ft.) above sea level, or more, but most of the population lives at elevations between 1000 and 1500 m. (3300 to 5000 ft.) This area between 1000 and 1500 m comprises a distinct ecological zone, quite different from the plains below and the cold windy peaks above, and labelled by its occupants "Shambaai" (Feierman 1974).

Because of their height the Usambaras are cool and wet, capturing moisture-laden air from the Indian Ocean. Three distinct rainy seasons are recognized: March to May, August, and November to December. Precipitation in the district headquarters town of Lushoto has averaged about 1100 millimeters (mm) (43 inches) per year over the past 50 years, and reaches 2,000 mm (80 inches) per year in the southeastern parts of the mountains. Annual temperature averages 18° C (64° F.), but substantial annual variation occurs, from August lows of freezing in the higher elevations to December highs of 29° C. (84° F).

Usambara is an area of high population density. In an area of approximately 3500 square kilometers (km²) (1350 square miles), Lushoto District contains over 275,000 people (see Fig. 1). The majority of the people are Shambaa agriculturalists, living in compact villages of 5 to 50 houses which are distributed fairly evenly

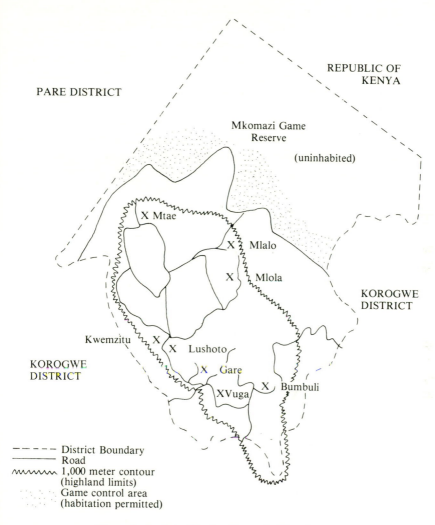

Figure 1. Lushoto District, Tanzania (Usambara)

throughout the highlands and linked by unimproved roads and
footpaths. The population is concentrated in the highland portions of
the district, where overall population density is at least 130 persons
per km² (Cliffe et al. 1975:149–151); in the areas around Soni and
Bumbuli (see Fig. 1) density is over 300 persons per km². As a
consequence, the size of individual farms is quite small. More than

half of all Shambaa farms are less than one half hectare in size (ca. one acre) (Attems 1967; P. Fleuret 1979). This landholding is generally divided into three to six small parcels, scattered at some distance from the home and from each other. The produce of this landholding supports a household of some five to six members. In the past Shambaa woman could inherit, purchase and dispose of land with some freedom, but rising population, rising land prices and changes in traditional inheritance rules have worked to concentrate land in the hands of men. At marriage a woman usually goes to reside with her husband in a house constructed adjacent to the husband's landholdings, and usually in the village of the husband's father. The wife is then allocated a portion of her husband's landholdings upon which it is her responsibility to produce the food that will be consumed by the household. The crops she grows may include maize, bananas, cassava, sweet potatoes, beans, white potatoes, squash and cocoyams. She may also sell a portion of this production in periodic market-places and use the cash thus acquired to purchase other household necessities, such as cooking oil, matches and soap. Men also cultivate food crops—coffee, tea, vegetables and deciduous fruits—the sale of which provides the bulk of household income. Hence both men and women engage in the cultivation of bananas in Usambara.

THE BANANA

It is generally agreed that the banana originated in Southeast Asia (Simmonds 1966; Greenway 1944). Numerous wild precursors and domesticated species are found there, including the two species —*Musa balbisiana* and *Musa acuminata*—which are commonly held to be the parent species of most of the varieties cultivated around the world today (Simmonds 1966, 1976)[2]. Although a distinction is frequently made between bananas, a sweet fruit eaten raw, and plantains, low in sugar and eaten cooked, this distinction is botanically not justified. The term "banana" is now preferred as a descriptor for the fruits of all Musa varieties. The term "plantain" will not be used further.

There are two major competing theories which attempt to account for diffusion of the banana to East Africa from Southeast Asia. The first and more commonly accepted one argues that the banana was transported from Malaysia to Madagascar in canoes by Malay immigrants (Greenway 1944; Simmonds 1966). From Madagascar the

Figure 2. Banana-Producing areas of East Africa

plant spread to the Mozambique coast and thence gradually north and west. The coastal areas and intermittent moist highlands adjacent to the lakes of East Africa (see Fig. 2) are more conducive to banana cultivation than to that of most cereal crops, so that diffusion of the banana was ecologically fostered in the area. It is in these areas, particularly Bukoba, Kilimanjaro and Buganda, that the banana is still the primary source of calories.

The second explanation of the spread of bananas to East Africa suggests that the plant reached Equatorial Africa from Malaysia via India, the Middle East and the Nile Valley. Some oral traditions suggest that the Baganda, for whom the banana is still the staple foodstuff, believe that the fruit reached them from the north, rather than the south (Rutishauser 1963; Greenway 1944). In any event, written sources report that bananas were grown along the Kenya coast in the thirteenth and fourteenth centuries A. D., and a Portuguese botanist commented in 1543 on the excellent bananas obtainable at Sofala in Mozambique (Greenway 1944).

Although the banana plant is generally referred to as a "tree" it is actually the world's largest herb. It is propagated vegetatively from corms. The corm material gives rise to a number of suckers which grow more or less concentrically around the parent plant. Each sucker produces one inflorescence which, after emergence, turns downwards and opens to reveal rows of flowers. The first rows of flowers are female, while the last rows are male. The fruits develop from the female flowers without pollination. The suckers can be, and frequently are, transplanted in order to establish new groves.

Bananas in East Africa generally produce fruit a year after planting, although there is some variability, due to altitude, variety and size of the planting material (Acland 1971:10). For best yields, bananas require well-drained, porous soils, warm weather conditions and well-distributed rainfall averaging around 4 inches per month and providing constant soil moisture. These conditions occur naturally in the southern portions of Uganda adjacent to Lake Victoria, the area of East Africa where bananas are of greatest dietary and cultural importance (cf. Rutishauser 1963). Acland (1971:9) estimates that 1.5 million acres in Uganda are planted in bananas.

In some other areas of East Africa, where rainfall may be inadequate or seasonal, the soil moisture is provided by cultivating the crop in drainage areas or on riverbanks or by irrigation. These areas include highland and lakeshore zones of Kenya and the well-watered highlands of Tanzania, particularly Bukoba, Kilimanjaro, Mbeya and Usambara (see Fig. 2).

Under conditions of manual cultivation, with little use of fertilizer and limited weeding and pruning, yields average six to eight tons per acre (13 to 18 metric tons/hectare) per year in East Africa (Acland

TABLE I

Nutritional composition of the banana (*Musa* spp.) per 100 g edible portion[1]		Representative nutritive value of Tanganyikan bananas per 100 g edible portion[2]
Energy kJ	423	437
Cals	88	91
Moisture (%)	77.0	70
Fat (gm)	0.1	0.2
Protein (gm)	1.5	1.3
Carbohydrate (gm)	20.6	21
Calcium (mgm)	9	4.6
Phosphorus (mgm)	21	47.1
Iron (mgm)	1.4	0.4
B-Carotene equiv. (micrograms)	120	500
Thiamine (mgm)	0.03	n.d
Riboflavin (mgm)	0.03	n.d
Niacin (mgm)	6	n.d
Ascorbic acid (mgm)	9	14

1. From: FAO-PHS Food Composition Table for Use in Africa, 1968.
2. From Raymond and Jojo 1940.
n.d = no data.

1971:14). With the use of fertilizer and proper spacing and irrigation, characteristic of banana husbandry in Kilimanjaro (von Clemm 1964), yields may reach 15 tons/acre/year (34 tons/hectare/year). Simmonds (1966:190) gives an estimate of up to 5 tons/acre/year (11 tons/hectare/year) for Ugandan bananas, while Attems (1967:79) suggests 5 tons/hectare/year (2 tons/acre/year) for Usambara bananas when intercropped. In places like Usambara, according to Allan (1965:161–166), 1/3 acre (0.13 hectare) *per capita* planted in bananas can easily yield all the bulk food (food with a high carbohydrate and calorie content) needed by a community.

The results of chemical analysis of the banana are shown in Table I.

An early study of Tanzanian bananas (Raymond and Jojo 1940) yielded somewhat variant results for nutritional content (see Table I).

The major nutritional drawback of the banana is its extremely low protein content per 1000 calories (Anthony et al. 1979:35); it is also a poor source of vitamins of the B group. These deficiencies are at least partially compensated for by the common practice in banana consuming communities of serving a sauce of meat, milk or legumes to accompany the banana main dish (Bennet et al. 1965; Rutishauser 1963).

BANANAS IN NINETEENTH-CENTURY USAMBARA

Very little archeological investigation has been carried out in northeastern Tanzania, so it is difficult to comment with any certainty about the human settlement of the Usambara Mountains or the origins of cultivation there. The historian Feierman (1974:23) argues that the Shambaa people of Usambara "... live today where they lived in the eighteenth century." What archeological evidence there is indicates the probability that Shambaa occupation of the Usambara Mountains is several hundred years old. It is uncertain when Shambaa people began to cultivate bananas, but Feierman assumes that the fruit has always been the staple food; he further argues that the substantial towns that served as administrative centers of the chiefdoms that comprised the pre-colonial Shambaa kingdom could have been supported only by the production of bananas (Feierman 1974:30).

Between 1848, when the Anglican missionary J. L. Krapf became the first European to visit the royal capital of the Shambaa kingdom at Vuga (see Fig 1), and 1890, when the first permanent resident Germans arrived, a fairly regular stream of missionaries and explorers visited the mountains, and all of them commented on the vast lush fields of bananas that were to be found there. When Sir Richard Burton passed through Vuga in 1857 it had, according to his estimate, 500 houses and was supported and provisioned by banana groves that circled the town (cf. Speke 1944). Writing in 1857, an Anglican missionary commented

> I think I never saw so many bananas as are growing all about Vuga. The people dry them and pound them into a sort of cake. They almost live upon them.[3]

In later German accounts the hillsides around the principal areas of settlement were said, typically, to be decked in forests of bananas (NOAM 1898:4). Holst, who made the first systematic study of Shambaa agriculture, estimated that up to 80% of all cultivated land was given over to banana groves (Warburg 1894:175). Yet another missionary (Smythies in 1894) commented that near Bumbuli "There is little cultivation, but great quantities of sugar cane and immense quantities of bananas. This is the staple food of the people ... very insipid tasteless food ...."[4] Baumann, writing of the interior of the mountains in 1891, states that one area had a population of 3000 and was characterized by "lush lawns and meadows" and "thick banana fields"; he describes bananas as the principal foodstuff. Given these and numerous other references, there seems little doubt that from at least 1800 on bananas were the single most important crop and food item in Usambara.

Indeed, Shambaa agriculture was designed specifically to foster banana production. As with the Chagga people of Mt. Kilimanjaro to the north (von Clemm 1964), the Shambaa irrigated some of their banana groves. Bananas were also planted in freshly-cleared earth and allowed to mature without manuring or irrigation; unlike maize, bananas were also often planted in previously-cultivated plots of ground, and new groves could be set out in tired maize fields. However, visitors to Usambara in the nineteenth century were uniformly impressed by a singular feature of highland agriculture: that is, a well-engineered and effective system of irrigation. Reservoirs and furrows captured and led water along hillside contours to fields of bananas, beans and maize. The missionary J. Erhardt, who visited Usambara in 1853, described his encounter with the ruler of the Shambaa kingdom, Kimweri:

> I found him seated ... under a large tree, close to a channel of water which is brought from a great distance for watering his women's plantations.

Erhardt wrote in his journal that Kimweri at this time maintained some 400–500 wives, probably an overstatement, but the amount of water required by an establishment even a tenth of that size would be considerable. Leaving the meeting place, Erhardt remarked on other irrigation works:

> I left ... descending the hill on which numerous reservoirs and water channels have been made for irrigating the king's banana and sugar cane plantations.

The use of irrigation was not, however, limited to royalty or to royal purposes. Baumann (1891:167), visiting the mountains in 1887–1888, noted a "rather developed irrigation system for the fields" in the Mlalo area and in the vicinity of Vuga saw "... everywhere small dammed-up canals, that watered the banana and bean fields of the slopes."

Bananas were also important as a trade item in the indigenous marketplaces which linked contiguous ecological zones into a system of exchange. The highest, coldest parts of the Usambaras were occupied by a pastoral people, the Wambugu, who produced milk, ghee and hides, while the plains regions surrounding the mountains provided millet, maize and salt; these goods were exchanged for commodities produced by the Shambaa, particularly bananas. Bananas were sufficiently important that they occupied a place in the local scale of values of marketable commodities—the only agricultural good, apart from beer brewed from sugar cane, to have a place in the scale.[5]

In 1850, Krapf wrote:

> There was today a *gulio* i.e. a market in the vicinity of Fairi. The natives brought in their articles of exchange, as: salt, bananas, fowls, butter, beads, cloth etc.

He also noted that Shambaa women carried bananas down to the plains market at Mombo. Descriptions of other traditional market-places at Malindi, Mlalo and Gare also point up the importance of bananas as a trade item. A useful description of the Malindi market has been left by a missionary's wife:

> It was market day and a great crowd of Wambugu met us on the road. Some flew from us like a hunted wild beast, when they saw from a distance our pack train approaching. Others greeted me ... others just stood dumb with astonishment. Each carried a huge cargo of maize or bananas into their home-woods (NOAM 1900(6):177).

Unfortunately, the early visitors to and residents of Usambara did not pursure the question of Shambaa nomenclature for bananas or their recognition of different varieties with much enthusiasm. Both Holst (1892) and Engler (1895), the major authorities on precontact Shambaa cultigens and agricultural practices, list only three named varieties of bananas in their writings. These varieties were called *maboko*, *mauti* and *mkono wa tembo*. Engler described only *mkono wa tembo* in detail, apparently more for its curiosity value (*mkono wa tembo* means "elephant arm or foot" and the fruits of this variety were supposed to be a foot in length) than out of scientific interest. Engler also noted the uses of portions of banana plants as roofing and packing material, string, twine and even pipe-stems. Hence, while we have useful objective accounts for the pre-colonial and colonial years of the cultivation, consumption and exchange of bananas, little subjective information about Shambaa classification and perception of the plants can be found. It is revealing, however, that the few names that Holst and Engler did collect correspond to the names of varieties cultivated presently and described by informants as being of great antiquity.

STAPLE FOOD CHANGE

After 1900, the primary position of the banana as staple food and principal cultigen in Usambara began to be eroded. Three other crops, two of them known to and cultivated by the Shambaa in precolonial times and the third introduced by the Germans, came to replace or supplement bananas in various parts of the mountains. The white potato (*Solanum tuberosum*), first planted by German missionaries near Mlalo in 1900, was accepted with enthusiasm by Shambaa farmers. There are two probable reasons for this. First, from a nutritional point of view, bananas and potatoes are closely comparable; calorie, vitamin and mineral content are approximately the same. There are also culinary parallels in that potatoes can be cooked, served and eaten in ways similar to steamed bananas or banana porridge. Second, the adoption of the potato permitted the expansion of Shambaa agriculture into higher, colder and drier zones of the mountains that were not congenial to banana production. When population growth in the highlands dictated the expansion of Shambaa agriculture, the potato, because of its dietary equivalence with the banana and its higher productivity when compared with other available cultigens suitable for cooler areas such as the sweet potato (*Ipomoea batatas*), was the natural choice.

Although maize was widely grown in Usambara at the time of European contact, it was much less important in the diets of most people than the banana. Although there are references to its cultivation and its role in trade in the nineteenth century, bananas were clearly much more important. However, after 1900, maize became increasingly more important, and by 1939 a district officer could comment:

> In certain parts of the Usambaras bananas where they are obtainable are the staple food, but, generally speaking, it may be said that maize is usually eaten in all parts of the district.[6]

Maize is much more labor-intensive and less productive (in terms of tons per hectare) than bananas. The explanation for the ascendancy of maize lies in its value as a cash crop. The Shambaa preferred to sell produce rather than perform labor as a means of earning cash, and the major markets for cash crops lay outside Usambara. Bananas were not suitable for this export trade, being bulky, difficult to transport, low in calories per unit weight and unappetizing to the lowland plantation workers they were supposed to feed.[7] Maize had none of these disadvantages. The necessity to sell food for cash to an exogenous market worked to reduce the attractiveness of banana cultivation.

Although bananas, once established, are not at all labor-intensive and produce impressive tonnages of food-with relatively little effort, the fruit's third and most recent competitor, cassava (*Manihot utilissima*), is even more productive and less demanding. A hectare of land planted in bananas will produce 11 million calories or more *per annum*, but the same land planted in cassava will produce between 20 and 27 million calories *per annum* (Fleuret and Fleuret 1980; Jones 1959:260). Population in Usambara is growing at a rate of 3.2% per year, population density exceeds 130 persons per km², and 54% of all farms are less than one half hectare in size. Given these circumstances, the superior productive capacity of cassava selects for its increased cultivation. Hence the greater feasibility of potato production in some areas, the greater economic returns of maize production, and the greater caloric returns of cassava production have all conspired to push bananas into a much less important role in diet and agriculture in contemporary Usambara. (For a more detailed discussion of these trends see P. Fleuret 1979; Fleuret and Fleuret 1980).

BANANAS IN PRESENT-DAY USAMBARA

Although the staple food role of bananas has been eclipsed by other agricultural commodities, banana groves are still quite conspicuous in the large areas of the mountains that satisfy their cultivation requirements. Bananas in Usambara are rarely grown in pure stands but are intercropped with one or more of the other basic foodstuffs. Bananas are also commonly planted in conjunction with coffee, one of the district's major cash crops, providing shade and mulch to the delicate coffee trees. Although the intercropping of coffee and bananas is not really beneficial to either crop, the practice is widespread in East Africa, including Buganda, Bukoba and Kilimanjaro (see Fig. 2).

Intercropping of bananas makes estimates of acreage devoted to them difficult, but a 1947 study of the Mlalo area estimated that banana plantings averaged 0.9 acre (0.36 hectare) per household of five to six members.[8] Residents of the banana-growing community of Kwemzitu owned an average of five farm plots per household in 1976, ranging from a few square meters to over a hectare (2.47 acres) in size, and grew bananas, intercropped with coffee, maize and beans, on an average of three of these plots. Banana production in Usambara and environs in 1975 was estimated at 52,000 tons (TIRDEP 1975). No figures were available on how much of this production was marketed and how much was consumed in village households. In their study of the marketing of cooking bananas in Dar es Salaam, Tanzania's largest city, Mascarenhas and Mbilinyi (1979) found that most of the fruits came from Morogoro Region; relatively few were from Tanga Region and none came specifically from Usambara. It seems clear that most Usambara bananas are consumed within Lushoto District.

Bananas play several important roles in contemporary Shambaa life. They are found as the main dish in about 3% of all meals sampled in a series of dietary surveys. These surveys were carried out in three different villages, each in a different micro-climatic zone of Usambara. Only one of the villages, Kwemzitu, produced substantial quantities of bananas; the second, Kwemtoni, produced maize and white potatoes and the third, Kwebaridi, maize and wheat. The results

TABLE II

Staple Food Consumption in Three Usambara Villages[a] in Tanzania
(meals per household)

Staple	Village Kwemzitu	Kwemtoni	Kwebaridi	Total
maize porridge	759	326	194	1249
cassava porridge	284	10	13	307
pounded maize	237	71	20	328
potatoes[1]	265	6	106	377
tubers[2]	24	21	2	47
bananas	46	—	27	73
rice	17	8	13	38
wheat	43	21	14	78

1. Includes both *Solanum tuberosum* and *Ipomoea batatas*.
2. Includes *Colocasia esculentum, Xanthosoma sagittifolium*, and *Manihot utilissima* roasted or boiled whole.
[a] The data were amalgamated from surveys conducted during the months of April, May, August, October and November, and January, reflecting an annual pattern of consumption.

Figure 3. *ng'ombe* banana variety, split for rack-drying, in a house yard

Figure 4. *ng'ombe* (1.) and *soo* (r.) banana varieties drying in a field (shamba)

Figure 5. *semkia* banana variety. Note lack of flower bud hanging from the stalk

of the surveys are summarized in Table II. Because maize is seasonal, but bananas are not, the consumption of bananas rises during the cool rainy months, especially March and April, when maize is in limited supply.

Cooked bananas, depending on variety, may be served as porridge, steamed paste or as part of a stew. Porridge is made from a flour of dried bananas. The bananas are peeled, split in half and placed to dry on a rock in the fields, on a rack built near the house or hung on pieces of twine suspended from the eaves of the house (see Fig 3–4). The flour is prepared by pounding the dried bananas in a mortar and pestle. Chunks of the stiff paste made from boiling the flour together with water are broken off with the fingers and dipped into a sauce of meat, beans or vegetables. Stewed bananas are usually fresh fruits with very firm flesh, peeled and cut into pieces and cooked together with beans, tomatoes, onions, oil or grated coconut. Steamed bananas are fresh fruits wrapped in banana leaf and steamed on a rock over the fire, or in a pot, until they become soft and pastelike. The resultant paste is eaten with the fingers after being dipped into a sauce, usually of beans or groundnuts.

More difficult to measure, but documented by conversation, child-following techniques (Wilson 1974; A. Fleuret 1979) and observation of village-level reciprocal transactions, is the snack or casual-food role of bananas. The fruits are consumed in large quantities by children on the path to school, on herding or firewood-collecting assignments and in the fields. It is not unusual to see a 12-year-old child eat three or four large fruits in less than half an hour. Adults are less likely to snack, but bananas do form an important component of inter-household food exchanges and are eaten by adults on festive occasions. Not only fruits, but suckers—seed materials—are exchanged among households. The principal basis for exchange is kinship, but the suckers are freely given to fellow villagers who are not kinsfolk as well. This practice has helped to foster the diffusion of promising new varieties of bananas among agricultural households.

Bananas also enter the Shambaa diet in alcoholic form—some varieties are manufactured into beer. This beverage is known as *mbege*. Bananas which have begun to go soft are pounded together with grain such as millet or maize hulls, cooked slowly for a day, then fermented for 24 hours. Fermentation is precipitated by adding the remnants of a previous batch or by the admixture of *Adansonia* (baobab) or *Kigelia* (sausage tree) fruits or yeast. The brewing, distribution and sale of *mbege* are primarily the responsibility of women who use fruits which they have cultivated in their own farm plots; they are entitled to keep the proceeds. The mineral content of *mbege* is quite high, particularly with regard to phosphorus and calcium; B-vitamin content may also be substantial (Acland 1971:15). Although ostensibly beer consumption is limited to men, women do drink beer on occasion, particularly during the rainy weather when agricultural work is at a minimum. Alcohol content of two samples from Bumbuli ranged between 2 and 3.5%.[9]

Numerous varieties of bananas are still very important in trade. Some of the administrative centers of Usambara support sizeable non-cultivating populations who must purchase all their food supplies. Many originate from banana-consuming areas such as Kilimanjaro; the larger markets within Usambara all exhibit steady and regular demand for bananas. Village women know of this demand and many cater to it, confident that a load of bananas carried to the market can always be sold. Of 57 women from the banana-growing village of Kwemzitu who were interviewed about their marketing activities, 35 named bananas as their first choice of an item to sell. One of these women who kept accounts of her marketing income averaged shs. 9/- ($1.15) per twice-weekly market day from her banana sales.

Assuming an average selling price of 10 cents per fruit, this means that if she sold all of the fruit that she brought, this woman transported approximately 90 bananas to the market, on foot. This income may be increased by travelling to markets in different ecological zones where particular varieties may be in demand. Kwemzitu women travel to the Mazinde market at the base of the Usambara escarpment, where individual fruits of certain sweet varieties may fetch 15 cents each, as opposed to 10 cents in a highland market. Women from Mlalo travel 25 miles (40 km) by bus to the twice-weekly marketplace in Lushoto town to sell a special kind of cooking banana which fetches 1/- shilling per fruit. Since they travel by bus rather than on foot, these women can bring very large quantities of fruit and earn substantial amounts of money, which is theirs to dispose of as they wish.

Bananas also serve significant non-food uses. Most important of these is the use of leaves as thatching material. Although well-to-do Shambaa farmers can now afford to purchase corrugated iron sheets for their roofs, the majority still use thatch. Banana leaf has the advantages of wide availability and low cost. Even when the leaf is dry and brown, it remains water-repellant. Many women prefer banana leaf to iron for kitchen roofs because the thatch "breathes" and permits smoke to escape. In addition to its use as thatch, banana fiber is also employed to make twine, in the manufacture of baskets for home food storage and for weaving sleeping mats. A fresh leaf makes a convenient umbrella. The leaves are also used as wrappings for cooking, steaming and storing food and for making wrappers for certain marketable commodities, particularly tobacco.

Although Engler and Holst only recorded three locally-named varieties of banana as being widespread in the Usambara of the 1890s, contemporary investigations in Kwemzitu, a sizeable village on the fringes of the area most suitable for banana cultivation, showed that nineteen different named varieties were being grown. The principal criteria upon which varieties are distinguished from one another include growing qualities (size of plant, time required from planting to harvesting, size of bunch); principal method of preparation (eaten fresh, cooked fresh, dried and made into flour, prepared as beer); qualities of the fruit such as size, color when ripe (red, green, yellow), other morphological characteristics; origin ("ancestral," coastal, European-introduced). Of these 19 varieties, 15 were found to be for sale in the large periodic market at the district headquarters in Lushoto town. Appendix A provides a listing of these 19 varieties and a brief description of each.

It is interesting that, despite the diminishing importance of the banana in Shambaa agriculture, a large number of varieties are maintained in cultivation and a good deal of indigenous experimentation continues to take place. Planting materials are continuously exchanged and new varieties are carefully assessed in an atmosphere of open inquiry.

CONCLUSION

In the world as a whole, bananas are considered a minor staple food, ranking far behind cereals such as maize, wheat, rice, sorghum and millet in total production. In those places where they are cultivated, however, bananas may play an extraordinarily important role, not only in the diet but in the domestic economy. We have attempted to illustrate this proposition by outlining the historical and current role of the banana in agriculture, diet and household economy among the Shambaa of Tanzania. Perhaps the most significant point relates to the decline of the banana within the staple food economy. In 1890 the banana was the predominant Shambaa staple. Two generations later, by 1940, it had been largely replaced by other staple foods. Two processes—the development of a regional market economy and population growth—were responsible for this change. By contrast, the introduction during the years 1920—1960 of commercial crops like coffee, tea, vegetables and deciduous fruits, although successful to various degrees, has had relatively little impact on the Shambaa staple food economy. In Usambara, at least, food production patterns are relatively resistant to changes consequent to the introduction of cash crops but appear to be quite sensitive to more general economic and demographic processes.

NOTES

1. The research on which this article is based was carried out in Lushoto District, Tanzania, December 1975 to February 1977. Support was provided by the National Science Foundation and the Social Science Research Council. Institutional affiliation was with the Department of Sociology, University of Dar es Salaam.
2. Modern cultivated bananas are distinguished from one another as varieties, not species (see Simmonds 1966, 1976).
3. From a manuscript contained in the archives of the Universities' Mission to Central Africa, London.
4. *Central Africa*, 1894, p. 19. This journal published letters from UMCA missionaries in the field.

5. According to LangHeinreich (NOAM 1907(2)), the scale of values from least to most valuable reads as follows: hen, flat basket of dried bananas, bundle of salt in banana leaves, the iron of a hoe, large basket of dried bananas, a jug of sugar cane beer, billygoat, female goat, female calf, bull, cow.

6. Lushoto District Annual Report, 1939.

7. Most migratory workers on sisal plantations near to Usambara came from the drier areas of western Tanzania where maize and millets were the principal foodstuffs. They were unfamiliar with bananas, did not know how to prepare them, and disliked the taste.

8. Mlalo Basin files, National Archives, Dar es Salaam. Total land-holdings per family were 3.43 acres (1.4 hectares), but due to interplanting a total of 6.50 acres (2.6 hectares) of different crops existed.

9. Analysis was done at Bumbuli Hospital by the Max Planck Nutrition Research Unit.

REFERENCES

Acland, Julien D. 1971. *East African Crops*. London: Longman

Allan, William. 1965. *The African Husbandman*. New York: Barnes and Noble

Anthony, Kenneth R. M., et al. 1979. *Agricultural Change in Tropical Africa*. Stanford, Calif.: Stanford University Press

Attems, Manfred G. 1967. *Bauernbetriebe in tropischen Höhenlagen Ostafrikas*. Munich: Weltforum Verlag

Baumann, Oskar. 1891. *Usambara und seine Nachbargebiete*. Berlin: D. Reiner

Bennett, F. J., et al. 1965. An inventory of Kiganda foods. *Uganda Journal* 29:45–54

Engler, Adolf. 1895. *Die Pflanzenwelt Ost-Afrikas und der Nachbargebiete*. Berlin: Geographische Verlagshandlung Dietrich Reimer

Feierman, Steven. 1974. *The Shambaa Kingdom*. Madison: University of Wisconsin

Fleuret, Anne. 1979. Methods for evaluation of the role of fruits and wild greens in Shambaa diet: A case study. *Medical Anthropology* 3:249–269

Fleuret, Patrick. 1979. *Farm and Market: A Study of Society and Agriculture in Tanzania*. Unpublished Ph.D. dissertation, University of California

Fleuret, Patrick and Anne Fleuret. 1980. Nutritional implications of staple food crop successions in Usambara, Tanzania. *Human Ecology* 8:311–327

Greenway, P. 1944. Origins of some East African food plants. *East African Agricultural Journal* 9:34–39, 115–119, 177–180.

Holst, B. 1892. Die Kulturen der Waschambaa. NOAM 1892(6):112–116

Jelliffe, Derrick B. 1966. *The Assessment of the Nutritional Status of the Community*. World Health Organization, Monograph Series, No. 53. Geneva: WHO

Jones, William O. 1959. *Manioc in Africa*. Stanford, Calif.: Stanford University Press

Krapf, Ludwig. 1860. *Travels, Researches and Missionary Labours during an Eighteen Years' Residence in Eastern Africa*. London: Trübner

Mascarenhas, A. and S. Mbilinyi. 1979. Bananas and the Dar es Salaam market. pp. 228–240 in *Papers on the Political Economy of Tanzania*, ed. by Kwan S. Kim, et al. Nairobi: Heinemann Educational Books

NOAM. 1887–1914. Nachrichten aus den Ost-Afrikanischen Mikssionen. Berlin.

Raymond, W. and W. Jojo. 1940. The nutritive value of some Tanganyika foodstuffs: I. The banana. *East African Agricultural Journal* 6:105–108

Rutishauser, I. 1963. Custom and child health in Buganda, IV: Food and nutrition. *Tropical and Geographical Medicine* 15:138–147

Simmonds, Norman W. 1966. *Bananas*. London: Longman

G

_____ 1976. *Evolution of Crop Plants*. London: Longman
Speke, John H. 1944. *What Led to the Discovery of the Source of the Nile*. London.
TIRDEP. 1975. *Tanga Integrated Rural Development Programme. Development Plan 1975–1980*. Tanga: Regional Development Director's Office
von Clemm, Michael. 1964. Agricultural productivity and sentiment on Kilimanjaro. Economic Botany 18:99–121
Warburg, Otto. 1894. *Die Kulturpflanzen Usambaras. Mittheilungen von Forschungsreisenden und Gelehrten aus den Deutschen Schutzgebieten*. Berlin: Ernst Siegfriend Mittler and Sohn
Wilson, Christine. 1974. Child-following: A technique for learning food and nutrient intakes. *Journal of Tropical Pediatrics and Environmental Child Health* 20:9–14

ADDITIONAL READING

Baker, R. and Norman W. Simmonds. 1951. Bananas in East Africa, Part I: The Botanical and Agricultural Status of the Crops. *Empire Journal of Experimental Agriculture* 19:283–290
_____ 1952. Bananas in East Africa, Part II: Annotated list of varieties. *Empire Journal of Experimental Agriculture* 20:66–76
Bogert, L. Jean. 1942. Dietary Uses of the Banana in Health and Disease: A Review of Scientific Literature, revised and enlarged edition. New York: United Fruit Company
Burden, O. and David Coursey. 1977. Bananas as a Food Crop. pp. 97–100 in *Food Crops of the Lowland Tropics* ed. by C. Leakey and J. Wills., Oxford: Oxford University Press
Cox, Paul A. 1980. Two Samoan Technologies for Breadfruit and Banana Preservation. *Economic Botany* 34:181–185
Dalziel, John M. 1937. *The Useful Plants of West Tropical Africa*. London: Crown Agents for the Colonies
Forsyth-Thompson, A. 1934. The Uses of the Banana. *Uganda Journal* 2:116–119
Friedrich, K. 1968. Coffee-Banana Holdings at Bukoba. pp. 175–212 in *Smallholder Farming and Smallholder Development in Tanzania*, ed. by Hans Ruthenberg. Munich: Weltforum Verlag für IFO Institut
Haarer, Alec E. 1964. *Modern Banana Production*. London: Hill
Haig, N. 1938. An Agricultural Survey in Buganda. *East African Agricultural Journal* 3:450–456.
Jameson, J. 1958. Protein Content of Subsistence Crops in Uganda. *East Afrcian Agricultural Journal* 24:67–69
Jameson, J. D. (ed.). 1970. *Agriculture in Uganda*. second edition. Oxford: Oxford University Press (see especially Ch. 11, "Staple Food Crops")
Kariki, S. 1972. Plantain Growing in Ghana. *World Crops* 24:22–24
Lucas, S. 1980. Consider the Banana. *The Bulletin, Pacific Tropical Botanical Garden* 10:58–62
Masefield, Geoffry. 1938. The Production of Native Beer in Uganda. *East African Agricultural Journal* 3:362–364
_____ 1944. Some Recent Observations on the Plantain Crop in Buganda. *East African Agricultural Journal* 10:12–17
_____ 1948. The Life of Perennial Crops. *East African Agricultural Journal* 14:160–161
Nagy, Steven and Philip E. Shaw (eds.). *Tropical and Subtropical Fruits: Composition, Properties and Uses*. Westport, Conn.: AVI Publishers
Palmer, J. 1979. Banana Products. pp. 625–635 in *Tropical Foods: Chemistry and*

Nutrition, vol. 2, ed. by George Inglett and George Charalambous. New York: Academic Press

Purseglove, John W. 1975. *Tropical Crops: Monocotyledons*. second impression. New York: John Wiley

Randiga, H. 1971. *The Banana Regions of East Africa: The Regional Distribution and Cultural Significance of a Traditional Food Crop*. M. A. Thesis, Geography, Western Kentucky University, Bowling Green

Reining, Priscilla. 1970. Social Factors and Food Production in an East African Peasant Society: The Haya. pp. 41–89 in *African Food Production Systems*, ed. by Peter McLoughlin. Baltimore: Johns Hopkins University Press

Ruthenberg, Hans. 1976. *Farming Systems in the Tropics*. second edition. Oxford: Clarendon Press (see especially Chapter 8, "Systems with Perennial Crops")

Simmonds, Norman W. and K. Shepherd. 1955. The Taxonomy and Origins of the Cultivated Bananas. *Journal of the Linneaean Society (London). Botany* 55:302–312

APPENDIX A

Banana varieties cultivated in Kwemzitu

Variety	Description
*tindii Ulaya**	The so-called copper banana, it has a dark red skin. It is sweet in flavor, eaten raw, and is said to have been brought by the Europeans (Ulaya-Europe)
*huuti**	The most important single variety. It may be eaten raw, but is more often sliced fresh and cooked in a sauce with oil (or coconut), salt, onion and beans. It is not generally dried. Corresponds to Engler's *mauti*.
*soo**	This variety is generally dried and then made into porridge. It has been grown for a very long time and is said to be the banana of the ancestors.
*tebwa**	A sweet variety, eaten fresh without cooking.
*kimalindi**	This variety comes from the coast of Tanzania and is prepared similarly to *huuti*. It was introduced in about 1950.
*libwi**	Dried and ground into flour for porridge.
*sheshekaa**	These may be eaten whole or dried to make porridge. They are similar to *ng'ombe*, but have very large individual fruits.
*katondwe**	The short, sweet, yellow finger bananas.
*ng'ombe**	This is another very common kind. They may be eaten whole or dried on a rock in the sun for flour and porridge.
*kalema**	Both the tree and the fruits are short. It is best eaten fresh. From the time a sucker is planted until fruit comes for this variety is one year, as opposed to eighteen months for many other varieties; hence it is much liked, despite the small fruits.
*mwema (mlema)**	A sweet banana, a larger variety of the *kalema* with a longer maturing time.
*kibiti**	A sweet banana similar to but larger than the *katondwe*.

	It is picked green, but allowed to turn yellow before it is eaten or sold.
*kijivu**	A sweet banana, yellow when ripe. It was brought from Uganda in about 1955.
*toki**	This is a brand new variety brought about five years ago (1975) from Moshi, Kilimanjaro. It is very popular; eaten cooked, fresh.
*semkia**	"No tail"—unlike others which have the flower bud hanging down from the stalk, these do not (see Fig. 5). They are for drying only.
bokoboko	Usually comes in very long bunches. It is best cooked fresh. It is found all over the mountains and has been around for a very long time. Corresponds to Engler's *maboko*.
mchuzi wa kunde	Dried, pounded, and made into porridge.
hoye	Grows well with much fertilizer. It can be eaten boiled, roasted, or dried and made into porridge. It is a new variety; a tree planted last year has already fruited and produced a sucker.
mkono wa tembo	Dried for porridge. The fruits are very large, as described by Engler.

*Indicates varieties on sale at the Lushoto marketplace in a survey conducted on March 13, 1980.

CHAPTER 8

READING THE TARO CARDS: EXPLAINING AGRICULTURAL CHANGE IN PALAU

MARY MCCUTCHEON
Directorate of International Activities
Smithsonian Institution
Washington, D.C.
20560

ABSTRACT

Over the last 25 years, taro (*Colocasia esculenta*) cultivation in the Palau islands of Micronesia has changed. Up until about 1960 it was generally grown in intensively cultivated swamps with multi-layered deposits of organic fertilizers, laborious land preparation of the mud, and canals with gates to regulate soil moisture. Before contact with Europeans, the population of Palau probably ranged between 20,000 and 50,000. Such intensive methods used in an island with limited fresh water swamp land would certainly seem justified.

By 1900, after a century of contact and epidemic diseases, only 4000 people remained. Every ethnographic reference from these years, however, indicates that taro cultivation continued to be highly intensive, though presumably less swamp land was in use.

This conservatism persisted until about 1960. Since then the population has grown rapidly. Unexpectedly, taro is now being grown on dry land or in swamps under swidden cultivation, while the old intensive taro swamps go dormant. These newer systems require less labor, but, after preparation, the yield is less per unit of land and the product is of lower quality.

To find the reasons for this paradox, we look first at changes in the social, economic and dietary role of taro in Palau. Then we consider what has happened to the role of women, the main laborers in taro patches. Finally, and perhaps at the heart of the explanation, we study U.S. administrative policy: its value for cultural preservation on the one hand, and its promotion of economic dependence on the other.

Rules of agricultural economics might best take into account these unquantifiable variables in explaining agricultural change, planning future development and predicting its consequence.

INTRODUCTION

Major cultural changes have occurred under the four different foreign administrations that have governed the Palau Islands in Micronesia. Along with the political system, religion and the economy, the method of growing taro (*Colocasia esculenta*) has also felt the pressure of change. Formerly grown almost exclusively under highly intensive culture in neatly maintained swamps, it has, in the last twenty years, become more commonly grown only intermittently in the same wet lands or as part of a crop rotation cycle in dry land gardens. The fact that taro cultivation has changed certainly comes as no surprise, but the direction in which it has shifted and the particular time at which this shift occurred forces us to re-examine certain theories on the development of agriculture.

THE LAW OF LEAST EFFORT

Ester Boserup argued in 1965 that population pressures tend to lead to agricultural intensification. More mouths to feed means that a greater yield must come from a fixed amount of land. In related work, Robert Netting (1969) showed, for a West African case, that agricultural disintensification is a rational response to decreased population densities. In general, fewer manhours of work are required to produce a given amount of food from extensively cultivated farms, so, to continue a labor intensive agricultural system when the population does not demand it is to expend unnecessary energy. These simple statements of probable human behavior appear not to be validated in the Palauan example where just the reverse interrelationship between population pressure and agricultural intensification has taken place.

Estimates for the pre-contact (18th century) population of Palau range between 8000 and 50,000. Roland and Maryanne Force (1972:4), while not expressing a preference for any of the alternative estimates, believe the pre-contact population may have been as low as 8000, a figure derived from questionable counts of house platforms. Different archeological evidence can be used, however, to argue for much higher estimates. Ancient people of Palau undertook a massive hillside terracing project which is still extant. Presumably these terraces were made to increase exploitable agricultural land for a large population. Between 1400 and 1700, this demographic pressure even forced habitation of the "dog-toothed" and inhospitable "rock islands" (Osborne 1966:461–462). John Useem's (1949:6) higher pre-contact population estimate of 50,000 may be defended by these data. A pre-contact population of 20,000, the lower value given by Useem, is now largely accepted as a reasonable compromise between the extremes.

Diseases ravaged the islands soon after contact and the population plummeted to less than 4000 by 1900. The number of people grew slowly over the next 30 years to 5794 in 1930. During the decade of the 1930s tens of thousands of Japanese civilians settled in Palau, but their stay was short lived. Because of the location of their settlements, this surge in population was of relatively small consequence for native taro cultivation. At the end of World War II, the number of Palauans was still under 6500 but was beginning to grow at an increasing rate. Even during those years of drastically reduced population, the labor and land intensive method of taro cultivation continued as before. In the last 30 years the population has sky-rocketed at a rate close to 4% a year to its present level of over 14,000. It is now, ironically, that disintensification of taro cultivation is taking place.

To find out why the law of least effort is not supported by the Palauan evidence, we must look at taro cultivation and taro itself in its cultural and historical framework. We must also investigate the purpose of taro cultivation in Palau. Not only has it been an important subsistence crop, but its production has important social meaning as well. My analysis of agricultural change in relation to the "purpose of production" follows the suggestions made by H. C. Brookfield in his paper "Intensification and Disintensification in Pacific Agriculture" (1972:37).

PURPOSES OF TARO PRODUCTION

Taro as a Food Source

Taro, in one form or another, fits into all Palauan dietary categories. When the leaf is cooked with coconut milk, the dish becomes an

TABLE I

Some Pre and Proto Contact Plant Foods of Palau

Common Name	Scientific Name
arrowroot	Tacca leontopetaloides
banana	Musa spp
mango	Mangifera indica
breadfruit	Artocarpus altilis
oriental mangrove	Bruguiera gymnorhiza
cycad	Cycas circinalis
tropical almond	Terminalia catappa
polynesian chestnut	Inocarpus fagifer
citrus fruit	Citrus spp
football fruit	Pangium edule
coconuts	Cocos nucifera
yams	Dioscorea spp
pandanus	Pandanus tectorius
nipa	Nipa fruticans
sweet potatoes	Ipomoea batata
swamp cabbage	Ipomoea aquatica
taro	Colocasia esculenta
taro, giant swamp	Cyrtosperma chamissonis
taro, wild	Alocasia macrorrhiza
mountain and wax apples	Eugenia spp
sugar cane	Saccharum officinarum

odoim, akin to our main course. The stem, peeled and boiled with sweetened coconut milk, can be classed as a *kliou*, more or less glossed as dessert. But the crop is best known for providing a starchy staple—an *ongraol*. The corm (the thick subterranean part) is boiled and peeled and then either eaten as is or prepared with coconut oil in a number of creative ways. It is known in Palauan as *kukau*. Until recently, kukau was unfailingly served at every meal (Barnett 1960:25), while all alternative starches were either unpopular, seasonal or undomesticated.

Other starches that were known in Palau, though seldom consumed, include yams, breadfruit, other kinds of taro, pandanus, oriental mangrove and sweet potatoes (see Table I). Long before European contact, yams had been introduced repeatedly from other Micronesian islands, but they failed to become popular (McKnight and Obak 1959:22, 29–30). Captain Wilson, who was shipwrecked with his crew of Englishmen in 1783, described a diet composed mainly of fish and "yams" (Keate 1788:49). The Palauan term Keate gives for these yams (*cocow*), as well as his discussion of their cultivation in swamp land, reveal that they were not true yams

(*Dioscorea*) at all, but certainly must have been taro. Breadfruit (*Artocarpus altilis*) is only seasonally productive in Palau. *Alocasia* taro and pandanus (*Pandanus tectorius*), both growing wild, are considered unpalatable. Other wild starches, such as oriental mangrove (*Bruguiera gymnorhiza*), grow in the mangrove swamps and are hard to obtain. Though starchy from a nutritionist's standpoint, arrowroot (*Tacca leontopetaloides*) belongs in the congitive food category of *kliou*. Sweet potatoes (*Ipomoea batatas*), probably introduced in the 17th or 18th century, were noted by Delano (1817:70) in 1791, by Holden (1836:57) in 1843 and by Cheyne (Shineberg 1971) in 1843; but they remained unpopular until this century. Except for the gardens planted by Yapese money miners, *Cyrtosperma chamissonis*, or giant swamp taro, appears to be a recent major food source. A native Palauan variety grows wild in interior swamps but is rarely eaten. Elderly Palauan informants are firm in asserting that only in this century has *Cyrtosperma* become a regular part of the repertoire of ongraols.

According to agricultural historians and to Palauan informants alike, almost all of today's year-round domesticated starches, except for *Colocasia* taro, were introduced in proto—or post-contact times. Until this century, there was a near total reliance on taro as ongraol, a category of food which must be present in every well-balanced Palauan meal. Its importance as a dietary staple cannot be under-estimated.

The Non-Subsistence Role of Taro

It is not surprising that such an indispensible staple crop also has important non-dietary significance. The number of varieties of taro which are cognitively and terminologically distinguished in the Palauan language is a clue to the vital role that it plays in the cultural system. McKnight and Obak (1960) counted hundreds of different labeled genetic clones, some of which have earned specific ceremonial functions. The distinctive features of a variety can be discerned only by close inspection of the most minute attributes. Cultivators who accept the significance of these types, however, can with little effort recognize many of them. Taro played a central part in the social, economic and ceremonial aspects of traditional Palauan life.

Taro is a symbol of womanhood and woman's role in the community. This association is a natural outgrowth of the division of labor. As is true in most of Southwest Oceania (Barrau 1965), the preparation, cultivation and harvesting of swamp-grown taro is solely the work of women. And, since taro held such a prominent position in the diet and economy of Palau, a woman's reputation was contingent on her dutiful production of this staple. Both the amount and the quality were yard sticks to measure her competence as a wife and as a woman.

Marriage in Palau, as everywhere, is partly an economic alliance (see Barnett 1949:124–127; Smith 1977:177–225). A woman's work and service as a wife meant big money to her brothers and mother's brothers. A woman who failed in this regard could not easily redeem herself in the eyes of the community, her husband's family or her own relatives (Smith 1977:141–144). Throughout the duration of a marriage, a prescribed exchange is kept up between the affinally related matrilineages. On special occasions the husband's clan donates money and/or land to the wife's relatives, who, in return, give food and services. No food gift would be complete without a generous portion of the finest taro. The givers would be derelict in meeting their obligations if they had not included this essential commodity. There were thus strong social pressures encouraging a woman to work hard in the hot sun, thigh-deep in the muddy taro patch.

The swamp is a woman's exclusive domain and, with many women working simultaneously in adjacent taro patches, it is the place where they can get together, gossip, weave baskets and tell jokes. Small shaded ramadas built on levees in the swamp serve as places for the women to rest. And after a day's work in the blazing sun and warm mud, they might have a refreshing bath together in the gravel-bottomed irrigation canals that criss-cross the swamp.

During post-partum rituals, the steam from boiling taro and herbs has a special significance which is consistent with taro's association with womanhood. Women maintain that sitting in a steam-filled enclosure has the property of tightening the skin and muscles and aids in readjustment to a non-pregnant state.

So basic is taro to the economy of Palau that it is used as the standard by which the value of the smaller pieces of Palauan money are measured. A small piece of money is worth ten baskets of taro and somewhat larger pieces are worth an integer multiple of this denomination (see Barnett 1949:43; Ritzenthaler 1954:18).

Taro is a firmly rooted element of Palauan culture. Its importance in the traditional diet, the social organization and the economy may elucidate some of the reasons that the methods of cultivating the crop respond to more than just the caloric needs of the populace.

METHODS OF TARO CULTIVATION

There are three cognitively distinct ways in which taro is grown in Palau. The *mesei* method refers to the intensive cultivation of fresh water swamps, the *dechel* is an extensively cultivated swamp and the *sers* is an upland dry farm. It is the relative increase and decrease in importance of these three methods that provokes investigation.

The Mesei Method

The highly intensive mesei method reflects the multi-faceted value of taro. Both a large initial investment of labor and continuous upkeep characterize these plots owned by a family or, more recently, an individual. Less than 2000 square feet (186 square m) in size, as many as 200 may coexist side by side in the swamp. McKnight and Obak (1960:28–37) described the system in depth. Here, only certain features will be discussed.

With an annual rainfall of 155 inches (3900 mm), swamps in Palau are plentifully, though variably, provided with water. To insure reliable moisture and to provide drainage, taro gardeners many years ago constructed elaborate mazes of masterfully engineered canals which divide up the whole swamp into manageable units. Gateways at intersections of the ditches can be lifted or dropped to hasten, slow or reroute water through the system. Meticulous attention is paid to keeping a proper moisture balance in the soil. Beside the ditches, a complex of walkways and bridges allows dry access for the women who tend their crops.

The work of these ancestors must be kept up if the mesei are to remain productive. About twice a year the water channels need to be cleaned. Falling leaves and the growth of grass and weeds can clog up the system and prevent the water from running smoothly. Ditch cleaning is a cooperative venture; everyone who has an interest in the swamp participates. Maintenance of the irrigation system, as well as the elevated pathways, requires a regular investment of labor.

Planting and harvesting of the taro is another labor intensive activity. No sooner is one crop harvested than preparation of the ground for the next crop begins. First, the woman churns the soil with her hands to aerate it and break up the clods. The mud is so soft she sinks up to her thighs in it. To fertilize and to prevent downward percolation of nutrients from the ultra-saturated soil, she introduces a layer of banana leaves or grasses at a depth of three feet (1 m) under the mud. These procedures require so much labor that it is likely to take several days to prepare a plot of a mere 500 square feet (46 square m).

Taro is always cloned to perpetuate the plethora of varieties that have such specific social and ceremonial significance. For this reason stalks of the recently harvested crop are usually replanted after drying in the sun for a few days. The propagation of taro entails poking a hole in the mud every 15 to 18 inches (38 to 46 cm) in a grid pattern and sticking the stalk in to a depth that gives it adequate support. The earth around the new taro plant must be pressed lightly to prevent it from falling down. A second layer of banana leaves or grass is then papered thoroughly over the entire surface as a fertilizer, weed inhibiter and moisture retainer.

It takes 8 to 10 months for the crop to mature. During this period, the moisture in the ground must be monitored and the channels kept clean. If it seems that in spite of the layer of surface mulch, weeds and swamp grasses are taking hold in the soil, they must be pulled out and a new layer of leaves and grass is blanketed over the decaying old one.

The plants are ready for harvest when the dark purple corm can be seen emerging from the surface of the mud. Then they are pulled, the corms are cut off and the cycle is repeated. Sometimes a mesei may be left unused for as long as a year. Short term fallowing should not exceed a year, however, for the taste of the next harvest will suffer and it will not technically qualify as true mesei kukau. One of the salient attributes of an ideal mesei is continuous use.

The fertilization, water control devices and uninterrupted use allow high production of taro per unit of land. From data on proceeds from commercial cultivation of taro, McKnight and Obak (1960:23) calculated that a 500 square foot (46 m²) mesei yields enough taro, at 5 cents a pound, to earn $35.00. These figures show an astonishing (and doubtful) yield of 140 pounds per 100 square feet per season (69 kg per 10 square m). Massal and Barrau (1955:18) indicate that an acre of taro plantation may produce anywhere from three to eight tons (2.7 to 3.6 metric tons). This equals 13.7 to 36.5 pounds per 100 square feet (6.7 to 17.8 kg per 10 m²). Bayliss-Smith (1977:337) estimates yield on Ontong Java to be 17,820,000 Calories per hectare. This, with the data from Massal and Barrau (1955:20) that 100 g of taro contain between 100 to 145 Cal. (418 to 605 J), allows an estimate of taro productivity there of between 33 and 39 pounds per 100 square feet (16 to 19 kg per 10 m²). By my own estimates, based on observations of density of plantings and mean weight of corms, a mesei might produce between 35 and 45 pounds per 100 square feet (between 17 and 22 kilograms per 10 m^2).

Because soil conditions are so well regulated, the consistency of the root is uniformly good, the taste subtly nutty, and the color a luscious shade of lavender. Very little of the flesh must be discarded during preparation; usually only the skin is thriftily peeled away. Mesei taro is universally acclaimed to be the finest.

The Dechel Method

A less intensive taro plantation is known as a dechel. Both men and women work together to prepare the land. They hack away the grasses, trees and bushes growing on swampy ground, burn them as well as possible under the prevailing damp and soggy conditions and drag away the residue. As is true for the mesei, only the women engage in planting and harvesting the crop. They obtain stalks from another recently harvested taro patch and insert these at irregular intervals of about two and a half feet into the unprepared mud. They cover the area haphazardly with leaves and ignore it for the 8 to 10 months it takes for the plants to mature.

The same swamp land can be re-used two or three consecutive times before the crop suffers from soil exhaustion. Once this occurs, it takes at least five years for a dechel plot left fallow to regain its value. During this time of disuse, the grass grows back and small trees have a chance to take root once again. A swamp can only be consecutively re-used within this period of soil re-generation if it is fertilized thoroughly by application of organic matter. If this is the case, it is well on its way to becoming a fully intensive mesei.

The dechel method produces much larger corms, sometimes twice the size and weight of mesei taro. Although the stalks are planted relatively far apart, a comparison between dechel and mesei plots both under cultivation would prove the dechel to be more productive. By my estimates using data on density of plantings and weight of produce, a dechel yields between 37 and 50 pounds per 100 square feet (between 18 and 24 kg per 10 m^2). The necessity for frequent fallowing, however, reduces the apparent benefits and it turns out to be less productive than a mesei when production is averaged over the fallow cycle. The maximum possible average annual yield of dechel taro is about 20 pounds per 100 square feet (10 kg per m^2) assuming the land is used for three consecutive cropping seasons (about 27 months) and left fallow for five years. This compares unfavorably to the maximum average annual yield from a mesei: 46 pounds per 100 square feet (22 kg per 10 m²) for repeated crops of nine-month duration.

For another reason, the size and weight of the product are deceptive measures of effective productivity. Although the corms are large, a lot of the flesh must be thrown away during preparation and consumption. There may be large crusty hollow places and spongy sections which are inedible. Finally, the taste of dechel taro is considered inferior to the mesei variety. A connoisseur of taro would have no trouble distinguishing between the two.

The Sers Method

The dry land gardens for cultivating taro are called sers. Some taro grows in backyard gardens, former house sites or wooded land, but it is in the rugged and largely unforested interior of the island of Babeldaob where the big plots of cultivated dry land lie.

The soil is lateritic, derived from volcanic tuffs and breccias and humus-rich top soil is shallow (Vessel and Simonson 1958:286, 291). The quality of the land for agriculture may be limited, but there is no dense complex of vegetation to clear. Until the late 1970s, the grasses that cover the land were kept in temporary abeyance by burning. On a dry and breezy day, a strategically placed match is all it took to clear a large area of land. More recently, however, farmers are recognizing that grass fires are destructive of soil nutrients and allow ash runoff during rain storms and wind. Clearing today is more commonly done with axes, sickles and hoes.

The gardener never has long to wait for a good downpour to soften the soil in the garden. It is then easy to prepare. The remains of grass and grass roots are raked away, the large clods of dirt are broken with a pick or three-pronged fork hoe and furrows 18 inches (46 cm) apart are constructed horizontally across any slope to keep run-off to a minimum.

Stalks of taro are obtained from a recently harvested mesei, dechel or sers and are placed against the furrow at about a 45 degree angle. The base of the stalk is covered with a mulch of leaves or the lemon grass (*Cymbopogon citratus*) that frequently forms a low garden hedge.

During the 8 to 10 months that the crop is growing, the garden may need to be weeded occasionally. Women often form informal cooperative work groups for this purpose. One week they help on one sers and the next on another. Some of the social life of the mesei is thus transferred to the dry gardens, even though the latter are not necessarily contiguous. Instead of working in separate mesei and socializing over lunch or an afternoon break, women work together in one sers and socialize while working.

It is desirable to situate dry gardens next to one another, not because of geographic necessity, as in the mesei, but because of the threat of rat infestation. If one cleared area borders another, a smaller section of the garden's perimeter is exposed to the tall grasses where rats lurk. This is why gardens tend to cluster side by side.

Where soil is poor, dry land taro is interplanted with cassava, sweet potatoes and pineapples. In house sites, clearings in the forest and partially wooded grasslands, topsoil is generally thicker and natural vegetation is denser. There the sers may also be planted with bananas, cucumbers, watermelons, peppers, *Xanthosoma* taro, papayas, sour sop, tobacco, green onions, areca palms, beans and a host of other fruits and vegetables. (See note re: Table I)

Insect infestation can be a problem, but it can be alleviated through careful management. After every harvest a different crop should replace the previous one, and root crops should not succeed one another. With this kind of crop rotation, species-specific insect predators never have a chance to take hold. My informants maintain that the primary function of crop rotation is insect control; soil exhaustion is mentioned less often. Nevertheless, this system of crop succession must contribute to the continued fertility of the earth, otherwise heavy with clay and lacking humus. Ideally, fallowing should not be necessary. One informant claims to have cultivated the same sers land continuously for 25 years, although this is not verified.

Corms from a dry garden are smaller than either those of the mesei or those from the dechel. They are spaced about 15 inches (38 cm) apart on rows 18 inches (46 cm) apart. The yield is approximately 30 to 35 pounds per 100 square (15 to 17 kg per 10 m^2) feet per nine-month season—the least of the three cultivation methods.

Taking into account the percentage of each corm that has to be thrown away in food preparation, the contrast grows. Inconsistencies in water supplies, poor soil, insects and general neglect all result in a crop riddled with crusty holes and mealy portions. A relatively high percentage of flesh is discarded with the peels. Furthermore, what remains is maligned—the texture is dry, the color pasty and the taste is bland. It is an inferior product and is held in contempt by old women who still take pride in their fine mesei taro.

There may have been a time before European contact when dry land taro cultivation predominated. McKnight and Obak (1960:7) derive the information about such a legendary "time of the gods" from living informants. From a pan-Palauan archeological survey, Osborne (1966:153) suggests that the extensive hill-slope terracing, which is still visible, was for the dry land cultivation of taro. Neither kind of evidence is conclusive. It is certain, though, that the mesei was the established, time-honored and most common cultivation method between the time of the first written descriptions in 1783 (Keate 1788) and the mid-1960s. Now the dechel and the sers appear to be replacing the mesei as the taro cultivation method increasingly chosen by younger women. It should also be mentioned that many women, to the horror of their grandmothers and in-laws, are opting against growing taro by any method.

Two problems need to be recognized. One is the apparent tenacity of mesei cultivation during the period of reduced population. The second is the degeneration of this method under the more recent American administration when populations have again been rising. This current trend becomes more enigmatic in view of two additional factors. First, authorities are professing a reverence for native tradition and its preservation. And second, the modern predominance of individual land tenure should, according to prevalent theory, promote maximum productivity per unit of land.

HISTORY

Pre-1960s

Even before direct European contact, the influence of the early explorers was felt in Palau. It is probable that two foreign crops found their way to this archipelago via inter-island communication as much as a century before Palau's official discovery. Sweet potatoes were introduced to the Philippines by the Spaniards in the 16th and 17th centuries. Tobacco came to the Philippines from a westerly direction in the early 17th century. From there, both crops eventually reached Palau. The Palauan word for the sweet potato, *chemuti*, is derived from a Nahuatl word, *camote*, adding strength to the argument for an American-Filipino-Micronesian route for the crop (Yen 1974:347).

The first European visit of any consequence to Palau occurred in 1783 when Captain Henry Wilson's ship, the Antelope, ran aground on a Palauan reef. From that time on, Palau was a stopping point for occasional ships from England, the United States, Germany and Spain. It was during these early contact years that dogs, pigs, goats, sheep, chickens, cinnamon, cloves, nutmeg and cats were introduced (Hockin 1803:15; Delano 1817:69).

There is no written evidence of the date at which the population began to decline. It is clear, though, that the epidemics of influenza and dysentery which were most probably responsible were sweeping the islands during the latter half of the 19th century. According to 1899 census figures (of doubtful accuracy), the extant population had sunk to a nadir of less than 4000 (Useem 1949:7).

For a very brief period around 1910, a religious movement hinted at a change in taro production. Increasing exposure to European things and ideas accompanying the German rule (1899–1914) demanded ideological reconciliation with Palauan tradition. One reaction found expression in the form of a prophet named Rdiall. His visionary philosophy stressed integration of new with old and adoption, sometimes totally out of context, of German customs. Rdiall's prophecies, cloaked in an aura of mysticism and the supernatural, won acceptance by a small band of followers. Donning white cotton clothes, erecting flag-poles and writing in a European-inspired script, they attempted to set the example for the rest of Palau.

Rdiall's predictions included a vision of a time when food resources would need expansion. Growing vegetables, including taro, in dry land was his answer to this future plight. Soon small dry land farms sprang up on the hillsides of Rdiall's home village of Ngkeklau (McKnight and Obak 1960:7). The influence of this prophet was short-lived, however. In 1910 the German government declared the movement subversive and banned practice of the cult altogether. By 1912 all remnants of dry land taro gardens were gone (Vidich 1949:83).

Toward the end of the German administration, on November 26, 1912, a typhoon hurled violent winds, surf and rain on Palau. The immediate damage to taro swamps was bad enough, but the secondary destruction from subsequent insect infestation and fungus proved devastating. A taro famine ensued which, according to the sources of the German ethnographer, August Krämer (1926:33), lasted until

1925. During this time of low taro production, people were compelled
to rely on wild alternative starches—scarce, hard to obtain and
unpalatable. To alleviate the hardship, a German agriculturalist
decided to import something starchy, fast growing and adapted to
tropical moist and acid soils. This was cassava (*Manihot esculenta*).
The cuttings readily took root and cassava soon became a second
popular starch, retaining its subsistence status even after the taro
swamps were free of disease.

Japan governed Palau from 1914 to 1944. During these years other
starches attained important positions in the diet. Sweet potatoes
finally earned belated acceptance after a long history as an object of
Palauan contempt. Trukese laborers in the phosphate mines of
Southern Palau consumed quantities of locally grown, though
seasonal, breadfruit. Palauans, having shunned it until then, learned
to appreciate it as an alternative starch. Another innovation was a dry
land form of taro, *Xanthosoma sagittifolium*, coming also from Truk
and ultimately from the West Indies. All of these introduced starchy
foods thrived on dry land and were grown in house yards or in larger
highland gardens along with casava, tobacco and bananas.

Maintenance of the new dry land farms must have meant dividing
already scarce labor between the two different and distant agricultural
domains. The pressure this imposed on women was relieved somewhat
by the willingness of men to share the burden of cultivating the sers.
But women were still faced with the bulk of responsibility for both
types of garden.

Two newly popular starches, *Cyrtosperma* taro and rice, were
adapted to wet land. The antiquity of *Cyrtosperma chamissonis* is
probably great in Palau, but old informants deny having eaten it much
in their youth. By the period of Japanese administration, however, it
was a common alternative starch. Even though it needs swamp land, it
was customarily planted along the borders of mesei, thus not
competing to any great degree with *Colocasia* taro.

As the number of Japanese colonists living in Palau grew during the
early 1930s, so did the quantity of rice. At first it was an imported
food, but then to some extent it became a locally produced crop. The
paddies in which rice grew, however, were not the same swamps that
had been used by Palauans for taro, and the land-holders and most of
the laborers were Japanese colonists foreseeing permanent settlement
in Palau. Few Palauans had access to rice as daily fare at that time,
and its introduction and fleeting prominence as local cultigen
probably had little effect on Palauan agricultural techniques and

dietary preferences during those years of Japanese occupation.

For several reasons, the disintensification of taro cultivation in the century and a half after contact would be predictable. The once exclusive dependence on *Colocasia* taro diminished as other starchy foods found acceptance. A reduced population meant a lower work force for the maintenance of labor intensive plots of taro, and those who survived may have been weakened by disease. Furthermore, as crops requiring different conditions and labor investment were introduced, this limited work effort had to be focused in a variety of economic endeavors in a variety of places. There was little impingement on the swamps where mesei were established; other wetlands crops, such as rice and *Cyrtosperma*, grew either in interior swamps or on the margins of mesei. There was every ecological reason, based on cost-benefit logic, to convert to low labor methods of cultivation.

Instead, it appears from the literature (Kramer 1926:133; Barnett 1949:4; McKnight and Obak 1960:28–37) and from informants' recollections that mesei taro cultivation, with its elaborate water control devices and deep mulching, continued to predominate through the early 1960s. The dechel remained uncommon and the sers was used mainly for experimenting with new taro strains (McKnight and Obak 1960:33).

Certainly the presence of water control devices in the mesei was a factor leading to their continued use. The initial labor in their construction had long since been invested, and it takes several years before the channels and gates completely grow over and become non-functional. There is no reason to abandon a still profitable investment. Maybe this is compounded by a general human attitude of stubborn conservatism. ("This is the way we've always done it.") Some of the lag between cause and effect in ecological models can thus be accounted for.

If one woman in a community chooses to keep her mesei active, then it is common courtesy for the others to continue, however reluctantly, to do their share of the communal maintenance. If the woman is old and of high social rank, this tendency would be particularly evident. The roughly 113-year-old chieftess of the Northern Confederacy of Palau, Ebil-Reklai Keringilianged, related the story of her final decision, only 15 years ago, to abandon her mesei. Every morning in the swamp she was joined by a group of apparently conscientious women who worked the adjacent patches. She assumed that they shared her feminine love of the taro swamps until she chanced to overhear two women mutter that they wished she would break her leg so that they could finally quit. She graciously retired the following season.

"Of tastes there is no disputing." This saying applies here, where all the "eco-logic" we can muster has no impact on dietary preferences. Mesei taro is better tasting than dechel or sers taro, and taro, until the 1960s, was the starch of choice among all the alternatives. Laboring long hours in a hot mesei is what it cost to have the most desirable food and the life to which Palauans had grown accustomed.

As long as one of a woman's major duties continued to be the contribution of starch, then the best of all starches, mesei taro symbolized her accomplishments as a woman. Her value as a wife was symbolized, as well, by mesei taro in the ritual exchange in which a husband's family gives money and/or land in return for foods and labor. Ecological reasons to change cultivation practices were thus counteracted by ideological reasons not to.

Post-1960s

During this century the population of Palau has steadily climbed. When the U.S. took over administration under a United Nations trusteeship agreement, it was about 6500. By 1964 it had reached about 10,000 and it is now fast approaching the 15,000 mark. Even in rural areas, the high birth rate more than compensates for migration to the cities. The contemporary demographic situation may, at last, be justifying the intensive agricultural system which had hung on for so long.

The United States' explicit administrative policy encourages the retention of tradition and this would have to include traditional systems of agriculture. Generated out of a post-war spirit of freedom and cultural relativism, the terms of America's trusteeship were non-interference with native cultures and customs. For almost twenty years those islands were largely left alone. A pristine precontact state was not to emerge, however, like some post-war phoenix out of the ashes of the American-Japanese conflict on Palau. Consequently U.S. money has in recent years been appropriated for the preservation or at least documentation of legends, songs, folk medicines, dances and wood carving. Other funds have gone toward constructing a men's house with attention to every traditional detail, to rebuilding historic sites and to teaching Palauan culture in school. Federal moneys have also been earmarked for rural development schemes and short-term village work programs in the hopes that these opportunities will help stem the flow of young people to the city.

Growing concern has been expressed in the last few years about the diets of elementary school children, and federal lunch program administrators bemoan the reliance on canned fish, canned "fricasee of chicken wings" and the ubiquitous rice. Efforts have been made to induce planners and cooks to provide native produce, specifically fresh fish, taro and cassava to the children. The rationale is both to upgrade the nutritional status of the lunches and to encourage productivity of these foods on a local level.

It is popular economic wisdom that when rights to use, buy, sell, develop and profit from land are held by a single individual for the duration of his or her life, and when these rights pass to his or her designated heirs, then the motivation to improve land for personal profit is strong. This theory is not supported in the Palauan example.

Between 1938 and 1941 the Japanese government in Palau prepared a land registry in which the majority of the native-owned land was registered as the property of individual men and women. This was a sharp break in a tradition which asserted that rights in use and alienation of land belonged to lineages and clans as corporations. Taro patches, once the property of the user's husband's clan, became in most case the user's individual property. After the war this predicament was left to the devices, however inadequate in a legal sense, of native traditional leaders. But land did not revert back to clans or lineages, and exponential growth in confusion has resulted from each generation's problems in determining inheritance. One thing seems certain: Individual land tenure, with its assumed rights in purchase and sale of land, is here to stay. Hearings on land titles are currently under way in much of Palau. Most plots of taro swamp land have now been assigned to individuals as real property in fee simple, thus resolving, for the time being, the particular and immediate problems concerning those parcels (McGrath 1971).

In spite of this individualization of tenure, during the last 15 years the mesei have increasingly been left to go fallow, dechel predominate, and sers are growing in popularity. Here again there must be some explanation for these facts.

The inspection tour of the United Nations in 1963 brought worldwide attention to the neglect that Micronesia had experienced since the United States assumed its trusteeship role in 1947. With this new

found awareness, decision-makers and taxpayers were outraged, and
—in what has become the stereotype of American policy—guiltily
increased the budget directed toward these little-known islands.

After three typhoons struck and demolished Palau between 1964
and 1967, administrators had a good opportunity to demonstrate their
benevolence. There was massive destruction of docks, buildings, and
public facilities. Relief funds were opened wide and the repair work
was undertaken. These projects brought dollars, not only to the
capital city of Koror, but also to the rural villages where just as much
damage had occurred. Paying jobs also became available for women
who were attracted to this chance to earn money.

Meanwhile the governmental bureaucracy was on the road to the
flamboyantly involuted state which characterizes it today. Over
52% of all wage jobs are government jobs, and of the 48% minority,
most are dependent in one way or another on the flood of federal
funds pouring into Palau. In one small rural village, there are 41
households and over 50 government positions. Though Palau is
currently negotiating a status of "free association" with the United
States, fiscal independence is nowhere in sight.

Imported foods, such as canned milk, tuna fish, mackeral, chicken,
Spam and cookies stock supermarket shelves; and dependence on
these non-native foods has grown as people move into the city, take
demanding jobs and earn the wherewithal to buy them. Cereals,
dehydrated potatoes and bread are new starches; but the most popular
of all is rice. Fifty-pound sacks are available in all food stores, and
when the ship fails to bring a new supply, there is a general feeling of
deprivation. Such heavy reliance is placed on this commodity that a
family of six can consume a 50-pound sack of rice in three weeks.
Kukau is now the second choice. Over and over I observed that even
when taro was available, already cooked and ready to be peeled, rice
would be laboriously washed and cooked, even at the risk of letting
the taro get moldy and go to waste.

Taro remains an important symbolic element in a customary feast,
but rice is usually served as well. Since it is the rice which is eaten and
the taro just as often thrown out to the pigs, there is no longer any
pressure to serve the high quality mesei variety. Dechel taro and even
the dry farm variety are now perfectly acceptable.

With wage labor available for women, the duties and expectations
of women have undergone radical shifts. Those working at wage jobs
are forgiven for not tending taro patches—a heretofore unthinkable
breach of role. Older ladies still look with approbation on a young

woman who cultivates taro, but there is little social sanction against those who choose not to.

CONCLUSION

From the foregoing discussion on the purposes of taro production and the chronological events that altered its role, changes in cultivation practices become understandable. In clarifying the factors which explain the shift from mesei to the less intensive dechel and sers methods, this study supplements some of the recent literature on agricultural decision-making under particular social and economic conditions.

From the time of early Palauan ethnographic commentary in the 1860s until about 1960, every reference to *Colocasia* taro cultivation in Palau emphasized reliance on a highly intensive method known as mesei. The traditional way of growing this basic food crop retained its importance, even under conditions commonly thought to precipitate agricultural extensification. The population during that 100-year period was relatively low, ranging between one fifth and one third of its estimated precontact size. Female labor was increasingly divided among several domains as more food crops were introduced, which also detracted from the near-total dependence on taro as a starchy food. Yet swamp lands available for taro cultivation were not diminished. Considering these factors alone, a shift to less intensive farming might have been expected. What must be taken into account, in addition, are taste, the nature of agricultural investment and the purposes of production.

Even with the introduction of other crops that could qualify as starch such as sweet potatoes and cassava, there was no replacement for taro, especially mesei-grown taro, as the ongraol of choice. The elaborate irrigation works that characterize the mesei were built long before the memories of any living informants, and thus constitute a pre-existing investment in the fields. This may account, as H. C. Brookfield (1972:35) argues, for what he calls "resistence to dis-intensification." Adult women continued to see their role defined, in part, by the quality and quantity of taro they produced. Under the customary social exchange rules, a woman's clan was obliged to give gifts of food to her husband's clan and stood to gain money or occasionally land in exchange. Taro of the finest mesei variety was invariably presented as proof of the wife's hard work. In addition, women socialized in their adjacent mesei and maintained their portions of the canals out of courtesy to anyone downstream. The cognitive value of this social production of the crop may not be measurable in terms of kilocalories. Brookfield notes that in these settings "inputs may be wildly uneconomic when measured by calorific returns, yet wholly reasonable when measured against social returns" (Brookfield 1972:38).

The trend away from intensive taro cultivation in the last 20 years has been gradual. It has happened at a time when population has been increasing and when verbalized administrative policies and land tenure patterns make it seem, at first glance, improbable. Events in Palau's recent history have, however, served to weaken the effect of some of the conservative forces seen in an earlier era. These events have led to such an overhaul in economic and demographic structure that mesei taro cultivation has become anachronistic.

A relative abundance of money now flows into Palau through wage labor opportunities located in the city of Koror, as well as through war claims compensations. Women have been accepted into the urban wage work force, and their neglect of traditional responsibilities in the taro patches is tolerated. Without regular maintenance and constant use, the irrigation systems will soon become irrevocably clogged with silt and vegetation. Furthermore, with access to money, with shipping lines serving Palau from the U.S. and the Far East, and with a large land-poor urban population, store bought foods are popular. Of the imported starches, rice is the primary choice. Rice is in fact so well liked that it is often preferred over taro grown in a mesei, even in rural settings.

The general rule relating agricultural intensification to population growth assumes that all other things are equal. The trouble is that all other things are seldom equal. One of the major factors assumed in the theory is that the universe of production, consumption and distribution of resources is a closed one. The islands of the Pacific are no longer the pristine cultural laboratories that they were once thought to be, and the ecological horizons within which Palauans participate go far beyond the edge of the reef. No rule of culture and behavior that requires a closed economic system could apply there.

Timothy Bayliss-Smith (1977:347–348) has considered this problem in his analysis of food production and energy-harnessing technology. He stresses the importance of manipulations of international social and economic networks in maximizing production of what he calls "de facto energy." On Palau, the U.S. administration has established exchange rates for labor, war-time land destruction and other non-edible commodities. These are easily converted by way of American currency to energy in the form of imported foods, petroleum fuels and batteries. As measured by energy-harnessing technology, the current Palauan tendency to extensify taro gardens could not be labeled "uneconomic." If we do condemn this as irrational, it could only be because the complex pyramid on which this economic resource chain stands appears precariously flimsy.

Another "unequal thing" that has interfered with the utility of classic economic laws in Palauan agriculture is the non-dietary meaning ascribed to taro and taro cultivation. Most economic anthropologists recognize the importance of cognitive value in utility maximization strategies. It would be good to integrate these non-quantifiable measures into a model of economic behavior. In addition to energy and material that usually form the dependent and independent variables in most equations of agricultural economic strategy, our humanized economic woman also considers emotive resources and seeks, as well, to maximize her soical network, self-esteem and pleasure.

REFERENCES

Barnett, Homer 1949. *Palauan Society*. Eugene: University of Oregon Press.
_____ 1960. *Being a Palauan*. New York: Holt, Rinehart, Winston.
Barrau, Jacques 1961. *Subsistence Agriculture in Polynesia and Micronesia*. Bernice P. Bishop Museum Bulletin No. 223. Honolulu.
_____ 1965. L'Humide et le Sec: An Essay on Ethno iological Adaptation to Contrastive Environments in the Indo-Pacific Area. *Journal of the Polynesian Society* 74:329–346.
Bayliss-Smith, Timothy 1977. Energy Use and Economic Development in Pacific Communities. pp. 317–359 in *Subsistence and Survival: Rural Ecology in the Pacific*. ed. by Bayliss-Smith and Feachem London: Academic Press.
Boserup, Ester 1965. *The Conditions of Agricultural Growth*. Chicago: Aldine. Publishing
Brookfield, H. C. 1972. Intensification and Disintensification in Pacific Agriculture. *Pacific Viewpoint*, 13:30–48.
Delano, Amasa 1817. *Narrative of Voyages and Travels in the Northern and Southern Hemispheres Comprising Three Voyages Round the World; Together with a Voyage of Survey and Discovery in the Pacific Ocean and Oriental Islands*. Boston: House.
Force, Roland and Maryanne Force 1972. *Just One House: A Description and Analysis of Kinship in the Palau Islands*. Bernice P. Bishop Museum Bulletin No. 235. Honolulu.
Hockin, Rev. John Pearce 1803. *Supplement to the Account of the Pelew Islands*. London: Nicol.
Holden, Horace 1836. *A Narrative of the Shipwreck, Captivity, and Sufferings of Horace Holden and Ben Nute Who Were Cast Away in the American Ship Mentor on*

the Pelew Islands in the Year 1832. Boston: Russell, Shattuck.

Keate, George 1788. *An Account of the Pelew Islands.* London: Nicol.

Kramer, August. 1926. Palau. in *Ergebniss der Sudsee-Expedition 1908–1910. II Ethnographie; B. Mikronesien Band 3, Teilband 3.* ed. by G. Thilenius Hamburg: Friederichsen. (Translated by anonymous)

Massal, Emile and Jacques Barrau 1955. Pacific Subsistence Crops: Taros. *South Pacific Commission, Quarterly Bulletin* 5(2):17–21.

McGrath, William R. 1971. Resolving the Land Dilemma. *Micronesian Reporter* 19:9–16

McKnight, Robert and Adalbert Obak 1959. Yam Cultivation in Palau, pp. 14–37 in *Yam Cultivation in the Trust Territory.* Anthropological Working Papers, No. 4, ed. by J. de Young; Guam: Trust Territory of the Pacific Islands

—— 1960. Taro Cultivation in Palau. pp. 1–47 in *Taro Cultivation Practices and Beliefs,* pt. 1. Anthropological Working Papers, No. 6, pp. 1–47; Guam: Trust Territory of the Pacific Islands.

Netting, Robert McC. 1969. Ecosystems in Process: A Comparative Study of Change in Two West African Societies. National Museum of Canada Bulletin No. 230. *Contributions to Anthropology: Ecological Essays.* pp. 102–112. Ottowa: National Museum of Canada

Osborne, Douglas 1966. *Archeology of the Palau Islands.* Bernice P. Bishop Museum Bulletin No. 230. Honolulu.

Ritzenthaler, Robert F. 1954. *Native Money of Palau.* Milwaukee Public Museum Publication No. 1. Milwaukee:

Shineberg, Dorothy (ed) 1971. *The Trading Voyages of Andrew Cheyne 1841–1844.* Canberra: Australian National University

Smith, DeVerne Reed 1977. *The Ties that Bind.* Unpublished PhD Dissertation. Bryn Mawr College, Bryn Mawr, Pennsylvania.

Useem, John 1949. *Report on Palau.* Coordinated Investigation of Micronesian Anthropology Report No. 21, Pacific Science Congress, Washington, D.C.

Vessel, A. V. and Roy Simonson 1958. Soils and Agriculture of the Palau Islands. *Pacific Science* 12:281–298.

Vidich, A. J. 1949. *Political Factionalism in Palau.* Coordinated Investigation of Micronesian Anthropology Report No. 23, Pacific Science Congress, Washington, D.C.

Yen, Douglas E. 1974. *The Sweet Potato and Oceania: An Essay in Ethnobotany.* Bernic P. Bishop Museum Bulletin No. 236. Honolulu.

ADDITIONAL READING

Alkire, William 1977. *An Introduction to the Peoples and Cultures of Micronesia.* Menlo Park, California: Cummings Publishing Co.

Gale, Roger 1979. *Americanization of Micronesia.* Washington, D.C.: University Press of America.

Mayo, Harold 1954. *Report on the Plant Relocation Survey and Agricultural History of the Palau Islands.* Saipan, M.I.: TTPI.

Nufer, Harold 1978. *Micronesia Under American Rule.* Hicksville, N.Y.: Exposition Press.

Palau Community Action Agency 1976. *A History of Palau, Volumes I, II, III.* Koror, Palau.

Trust Territory of the Pacific Islands 1968. *A Guide to Subsistence Agriculture in Micronesia.* Agricultural Extension Bulletin #9. Saipan, M.I.: TTPI

CHAPTER 9

CASSAVA AND CHANGE IN PACIFIC ISLAND FOOD SYSTEMS

RANDOLPH R. THAMAN
School of Social and Economic Development
University of the South Pacific Suva, Fiji
AND
PAMELA M. THOMAS
Dept. of Geography
Australian National University
Canberra City, ACT
Australia

ABSTRACT

Cassava, a native of tropical America, has invaded the Pacific Islands. It has displaced traditional tuber crops such as yams and taro as the most important staple of many island groups. Although commonly seen as a major factor in the breakdown of traditional food systems and in increasing erosion and soil deterioration, as well as being considered of inferior nutritional status, cassava also has a number of positive characteristics. It is an excellent source of dietary fiber, energy and some micronutrients, and is one of the highest energy yielding crops, both in yield per unit area and energy output per fossil fuel input. It has an extremely wide environomental tolerance. Although susceptible to wind damage, it is very drought resistant, thrives in soils too poor to support other crops and can grow from sea-level to over 2000 meters. It is an excellent livestock feed, can be intercropped with a wide range of cash and subsistence crops and is virtually pest-free in the Pacific Islands.

The favorable and unfavorable characteristics of this remarkable plant, which is responsible for feeding approximately two thirds of the world's some 500 million people who subsist on root crops, are discussed. If optimally integrated into Pacific island food and agricultural systems, cassava could possibly play a major role in reversing dangerous trends of rapidly increasing rates of malnutrition and nutrition-related degenerative disease, as well as helping to ease economic crises related to dangerously high levels of food dependency in many island groups.

INTRODUCTION

In the Pacific Islands and throughout the Third World, processes associated with modernization have been responsible, either directly or indirectly for drastic changes in traditional food systems. These processes include changes in traditional economic systems and social organization resulting from urbanization; emphasis on cash-cropping for export; increasing desire for wage employment; labor and brain drain from rural areas; increasing importance of leisure and recreational activities; increasing population and pressure on land resources; and general environomental deterioration. Phenomena related to modernization have forced societies to adapt to these changing cultural and environmental parameters. Among Pacific Island societies a common strategy has been to rely increasingly on cassava as the staple food crop. The speed and the magnitude of this change is such that it has been referred to as the "cassava invasion" (Thaman and Thomas 1980, 1982).

Is cassava an appropriate technology for the Pacific Islands? What have been the cultural, nutritional and ecological impacts of its invasion of the Pacific Islands? How has it helped Islanders adapt to change? These are the major questions addressed in our discussion of whether cassava is or could be detrimental or advantageous to Pacific Islanders.

In examining the "cassava invasion" and its implications for the Pacific Islands, several aspects are examined: The general characteristics of cassava; the introduction, distribution and relative importance of cassava in the insular Pacific; the specific cultural, nutritional and environmental impacts it has had or could have on Pacific Island food systems. The ways in which increasing dependence on cassava has helped Pacific Island societies adapt to change. And finally, development proposals for optimum integration of cassava into Pacific Island cultural and environmental systems. Although this paper focuses on the Pacific Islands, an overview of cassava in other areas of the tropics is included in order to understand its potential impact in the study area.

CHARACTERISTICS OF CASSAVA

Although the European discovery of America in 1492 destroyed much of the existing social pattern, it had the beneficial side effect of disseminating New World crops to the Old and vice versa. ... of the food crops, perhaps the most important was cassava, whose cultivation has expanded almost explosively across the tropics in the last few decades. (Coursey 1978:137)

General Characteristics

A native of Mexico, Guatemala and northeast Brazil (Purseglove 1968:173), cassava, also called manioc or tapioca (*Manihot esculenta* Crantz), is the world's most important root crop. It is responsible for feeding approximately two thirds of the world's estimated 500 million people who subsist on root crops (Coursey and Booth 1977a:100; Coursey 1978:132); perhaps one billion people eat it (Nestel 1977:1). Since its introduction into the Pacific Islands in the nineteenth century, it has become an increasingly prominent staple in food systems where root crops are still synonomous with food and culture.

Cassava was introduced into the Pacific as a crop for flour and starch, for famine relief following hurricanes and, in some cases, for the direct consumption of its tubers. The characteristics of cassava clearly justify its dramatic rise in importance as a staple food crop. No other crop gives such substantial output for such minimal input. "It is one of the most efficient producers of carbohydrate known" (Coursey 1978:132) and "can be grown in poor soil ... reaped for little labor and kept in the ground until required" (Purseglove 1968:178).

Its ease of cultivation, environmental tolerance and high productivity make cassava synonymous with change in much of the tropical world. With food problems resulting from increased cash cropping, urban migration, degraded soils, shortened fallow periods, land shortages, expansion of grasslands and savannas at the expense of forest and increasing desertification (especially in Africa), many Third World countries have come up with the same answer—cassava. Its utility is overcoming its lack of status. While many Pacific people consider it to be "poor man's food" or "pig food" and provide yams, taro, sweet potatoes or rice for chiefs and other dignitaries, cassava nevertheless forms an increasingly large part of their own daily diets.

Cassava is a single species in the family Euphorbiaceae. Although there are two distinct types, bitter and sweet which are often classified as *Manihot esculenta* Crantz and *M. utilissima* Pohl respectively, they are now consider variants of the same species *M. esculenta*

(Coursey and Booth 1977b:77). There are a vast number of cultivars which exist wherever the crop is grown. This tremendous diversity is one reason for cassava's wide environmental tolerance, but it also leads to considerable difficulty in classification. The International Center for Tropical Agriculture (CIAT) in Colombia, for example, has a large germ plasm collection of between 2000 and 2300 cultivars and produces about 40,000 hybrid seedlings per year (Nestel 1977:2).

One significant characteristic of cassava is the presence of potentially toxic cyanogenic glucosides in the tubers and leaves. These glucosides, which hydrolyse to produce prussic acid, are found in much higher concentration in the bitter varieties. In the sweet cassava "the glucoside is largely confined to the peel and is at a low level" (Onwueme 1978:109). The concentration of prussic acid ranges from 10–490 mg per kg of tuber and tends to vary not only with the cultivar but with environment (Onwueme 1978:146). Drought, low soil fertility, and especially potassium deficiency or high nitrogen level of soils, reportedly increase root cyanogenesis (Coursey and Booth 1977b:77). Even with the most careful detoxification, the final cassava products still contain some prussic acid. The specific effects on humans and domesticated animals are discussed below.

Growth Characteristics

Of all the tropical staple food crops, cassava is the easiest to grow and requires the least labor. "Cassava gives good yields even on heavily cropped soils and does not require a lengthy fallow and also yields more per man-hour of labor than any other food crop" (Maude 1973:175). It produces well with little ground preparation and minimal subsequent attention. Cassava is a perennial propagated by thick stem cuttings which are widely available, as they are not the edible part of the plant.

While cassava grows well in low fertility soils, it needs relatively high concentrations of potassium and good drainage to produce large tubers and, as a result, does well where slash and burn methods, which yield high amounts of potash, are used. In high fertility soils it tends to produce excessive vegetation at the expense of tuber growth. Although weeding assists early growth, after two to three months there is often sufficient canopy to make further weeding unnecessary. Tuber formation commences by the second month after planting and, unless there is drought, it continues bulking until maturity (Onwueme 1978:110).

In the Pacific Islands the relatively brittle branches of cassava are considerably more susceptible to windfall and hurricane damage than traditional crops such as yam and taro. Once the stems are separated from the tuber, deterioration of the tuber is quite rapid.

Short-season cultivars (usually the sweet varieties) can mature as early as six months after planting, but between six and nine months is most common. These varieties, however, do not keep in the ground as well as many of the bitter cassavas which take between 12–18 months to mature and may be left in the ground three to four years without serious deterioration (Purseglove 1968:173). Cassava can be planted at any time, but where there is seasonal variation it does best when planted just before the rainy season.

Comparative Yields in Tropical Areas

In 1978 the estimated world area planted in cassava was 13,132,000 hectares and the world production 119,374,000 metric tons, an increase of 19.5 million tons over the 1971 production figure. Similarly, between 1971 and 1978 the production of cassava in the Pacific expanded from 17,000 hectares to 20,000 hectares (FAO 1979:116).

Yields vary considerably, depending upon a number of variables. "In India yields vary from 2.5 to 32 tonnes per hectare, while under intensive cultivation 50 to 65 tonnes per hectare have been recorded" (Purseglove 1968:178). On a world basis the estimated average yield is about 10 tonnes per hectare, with yields in South America averaging over 14 tonnes per hectare, while those in Africa are only about half as much (Onwueme 1978:135). According to Massal and Barrau (1956:23), "in the semi-foraging conditions prevailing in native gardens in the South Pacific, it cannot usually be expected to exceed between 2.5 to 10 tonnes per hectare, i.e. from 2,500,000 to 10,000,000 calories,"

United Nations Food and Agriculture Organization (FAO) figures for average cassava yields in selected tropical countries given in Table I indicate very high yields for some Pacific Island countries, especially the Cook Islands and New Caledonia. The figures are probably too low for Tonga. Either the two former countries are especially suited for cassava cultivation by virtue of both their physical and cultural

TABLE I

Average per hectare yields of Cassava in selected tropical and Pacific Island countries for 1978[a]

Country	Tonnes per hectare
Ghana	7.4
Nigeria	9.7
Indonesia	9.2
Brazil	11.5
India	16.7
Cook Islands	32.3
New Caledonia	25.0
French Polynesia	18.5
Fiji	12.3
Papua New Guinea	10.7
Tonga	5.9
Niue	4.0

[a] Source: FAO 1979:116–117.

TABLE II

Relative edible food production for various staple crops in tonnes and Gigajoules per hectare[a]

Crop	Tonnes/ha	GJ/ha
Cassava	9	50.2
Irish Potato	10	25.1
Yam	8	29.3
Sweet Potato	7	29.3
Taro	6	25.1
Rice	2	20.9
Maize	2	33.4
Sorghum	1	12.5
Wheat	1	16.7

[a] Source: Adapted from Onwueme 1978:156.

environments, or the statistics may be highly unreliable. Niue, with very poor soils and a devastating drought which extended into 1978, may have experienced very low yields. The same drought also affected Tonga. Measured in edible food calories per hectare, cassava produces more calories than white potato, yam, sweet potato, taro, rice, maize, sorghum or wheat (Table II).

The yield potential becomes even more dramatic when the yields per input of fossil fuel (a commodity of ever increasing cost and scarcity) are considered. Studies in Africa showed yields of 1164.8 Cal (4869 KJ) of food energy were obtained for every input of one Cal of fossil fuel, compared with figures of 128.8 Cal (538 KJ) for corn in Mexico; 37.5 Cal (157 KJ) for sorghum in Sudan; 3.91 Cal (16 KJ) for rice in the Philippines; and 1.08 Cal (4.5 KJ) for wheat in India (Pimentel 1979:81–82).

Moreover, there are reportedly cassava cultivars which out-yield traditional ones by 100% or more and have higher protein content or shorter growth periods (Norse 1979:46). In Latin America where cassava yields are consistently high, the development of new cassava technologies are aimed at further increasing productivity (Lynam 1978:213).

Environmental Tolerance

Cassava produces these high yields in a wide variety of climatic, environmental and soil conditions. Although it does best in the lowland tropics with a moist, warm climate and well distributed rainfall, "it will grow adequately up to 5000 feet (1524 m) on the equator and can cope with temperatures down to 10° C" (Purseglove 1968:174).[1] Bourke (1982:125) reports that in Papua New Guinea it is grown as a minor crop up to 1600 m (5240 feet) and has been recorded as high as 2100 m (6888 feet).

Cassava is extremely drought resistant and can be profitably grown within an annual rainfall variation of 50 to 500 cm (18–180 inches) (Purseglove 1968:1974). In drought conditions it merely stops growing and loses its leaves but recovers quickly after rain. "This behaviour makes cassava a valuable crop in places where rainfall is low, or uncertain, or both" (Onwueme 1978:110).

Cassava does best in light sandy soil with good drainage but can be grown on almost all soil types provided they are not waterlogged, too shallow or stony. "Cassava will produce an economic crop on exhausted soils unsuitable for other production and consequently is often the last crop taken in the rotation in shifting cultivation" (Purseglove 1968:174). Throughout the Pacific it is commonly grown where other crops will not produce economic yields. Among the Maring speaking people of the Bismarck Mountains of Papua New Guinea, cassava is planted "where the ground being hard was judged unsuitable for taro" (Clarke 1971:78). It is frequently planted in Fiji

H

on land considered too steep or too rocky for other crops. In Niue it is grown on the "poorer soils of the central basin" (Sykes 1970:94). While cassava is considered not to like saline soils, it nevertheless produces profitable crops on the more saline soils in many areas of the Pacific. Howlett (1967:54) states that it is a staple crop on the lowland saline soils of Papua New Guinea. "Poor fertility of many soils in the Pacific restricts the yield of traditional tuber-bearing plants and partly explains the development of cassava cultivation" (Massal and Barrau 1956:22). "Together these factors make cassava a low-risk crop highly adapted to more marginal agricultural areas" (Lynam 1978:220).

Effect on Soils

There appears to be some disagreement as to whether or not cassava is responsible of further degradation of already exhausted soils. "The expansion of cassava growing on some islands of the Pacific has resulted in serious soil degradation as it grows even on poor soils. The islanders have too great a tendency to exhaust them by almost continual cropping" (Massal and Barrau 1956:23). However, cassava produces a considerable quantity of leaf litter as plants drop their leaves every one to three months (Martin and Ruberté 1975:28), thus returning part of the nutrients which they contain to the soil (Onwueme 1978:132). Given cassava's greater ability than most plants to extract nutrients from the soil coupled with the partial return of nutrients in the leaves to the soil, cassava is in fact not as exhaustive to soils as is sometimes asserted. Its survival on poor soils "seems to reflect a plant with superb equipment for extracting nutrients. . . . an examination of this mechanism could provide useful information about the efficient utilization of soil nutrients" (Haynes 1977:45). This is a significant subject at a time of increasing fertilizer costs and other inputs.

Trials to ascertain the amount of nutrients removed from the soil in the harvest of various tropical crops show quite conclusively that, by weight of yield, cassava uses less nutrients than other staple crops (Table III). It is a common belief in Tonga that cassava, as a component of the fallow vegetation, is an indicator of high soil fertility. Maude (1965:136) cites an interesting example from the low lying coral-limestone islands of Tonga's Ha'apai group,

TABLE III

Nutrients removed in harvest per 454.5 kgs (1000 lbs) yield of specified crop (assuming all of the residues are left on the land)[a]

Crops	N	Kilograms per hectare		
		P	K	Ca
Maize (grain only)	16.8	3.0	3.0	0.22
Rice (paddy)	13.4	3.6	3.9	0.9
Peanuts (kernels and shells)	43.3	3.8	6.8	1.5
Cassava (fresh tubers,	2.5	0.34	3.7 to	0.30 to
30% dry matter)			6.5	0.61
Yam (fresh tubers,				
30% dry matter)	3.8	0.30	3.9	0.07
Bananas (fruit, 30% dry				
matter)	2.5	0.46	6.1	0.07
Cocoa (beans and husk)	24.7	4.5	35.9	

[a] Source: Adapted from Nye and Greenland 1969:28.

... where land has to be worked fairly intensively, at least by bush fallowing standards. Here, land under cassava is often not kept weeded but the secondary growth of the fallow (mainly grasses and other herbaceous plants) is allowed to spring up around the crops, which is [sic] only weeded as part of the cassava is harvested and replanted. The result may be described as a sort of combined cropping and fallow period, and farmers claim that both the secondary growth and the cassava help to restore the fertility of the soil and that when the area is finally cleared and harvested (after perhaps two years) it can be immediately planted in yams.

Cassava could in fact be less responsible for further degradation of soils than many other crops. Haynes (1977:45) suggests that it is only because it is so frequently used as the last crop in a rotation that "circumstances have given rise to the view that manioc depletes soil of nutrients."

Cassava can be intercropped with other annuals or tree crops and, in the Pacific, is frequently planted along the edges of garden patches (Clarke 1971:75) or combined with yams (Cassava tubers spread more widely and penetrate less deeply than yam). It is often intercropped with coconuts in Tonga and Fiji and with sweet potato and pineapple in Tonga (Thaman 1976; 1977a). It is also commonly planted with coconuts in India, Sri Lanka, Malaysia, Philippines and Brazil. In Tanzania coconuts intercropped with cassava had a higher net yield than those left uncultivated or around which the grass was cut occasionally (Gallasch 1976:104).

Susceptibility to Pests and Diseases

Within the Pacific area cassava is virtually pest and disease free. Elsewhere, its resistance to locust attacks as well as drought ensured its importance as an African "famine crop." Bitter cassava has the added African advantages of being unpalatable to baboon and hippopotamus. While hippoptami are unlikely to be a problem in the islands, a point about animal predation is relevant to the Pacific, where animal damage can be a problem. On the island of Mai'ao in Tahiti where cassava was once the staple crop, one reason for its being abandoned was that farmers were "fed up with having pigs eat their manioc. . . . it was a constant source of irritation and friction (Finney 1973:59). Bourke (1982:126) states that in Papua New Guinea "pig or rat damage to tubers can be significant." A change from sweet to bitter cassava, however, is not a feasible solution due to the problem of higher toxicity to both animals and humans and more difficult food preparation.

In Papua New Guinea cassava is susceptible to a "leaf spot caused by *Cercospora henningsii* and a disease caused by *Corticium salmonicolor* has been recorded as serious" (Bourke 1982:126). In Africa, India and Indonesia there are several other diseases affecting cassava, the most serious being the African cassava mosaic virus which causes leaf distortion, stunted plant growth and considerable reduction in tuber yield and quality (Onwueme 1978:139). Brown streak and several other fungal diseases have recently been recognized in Thailand, Malaysia and Taiwan. Cassava is also susceptible to vascular wilt which occurs under conditions of excessive humidity (Williams 1975:156).

Nutritional Characteristics

> How come that in the story of a world conqueror like cassava, so much attention has been paid to the calorie stream from the roots and so little to the qualities of the leaves? (Oomen and Grubben 1978:11)

As Oomen and Grubben stress, the lack of protein in cassava is constantly noted, but mention is seldom made of the very high nutritional value of cassava leaves (Table IV). "Two cassava leaves contain sufficient carotene for one child and after cooking it becomes just one tablespoonful" (Oomen and Grubben 1978:124). The leaves are also relatively high in protein content (5 to 7% on a dry weight

TABLE IV

Composition per 100 g edible portion of selected Pacific Island vegetables[a]

Species	Dry matter	MJ	Protein g	Fiber g	Calcium mg	Iron (mg)	Carotene mg	Ascorbic Acid (mg)	Folate mcg
Cassava leaves (*Manihot esculenta*)	19.0	251	6.9	2.1	145	2.8	8.3	80	?
Taro leaves (*Colocasia esculenta*)	18.6	255	4.1	1.2	160	1.0	5.5	65	160
Amaranth (*Amaranthus sp.*)	15.2	180	5.2	1.0	340	4.1	7.7	120	85
Sweet Potato (*Ipomoea batatas*)	13.3	176	3.2	1.6	85	4.5	8.7	20	90
Pumpkin (*Cucurbita moschata*)	7.4	88	3.0	?	40	2.1	1.9	10	?
Chinese Cabbage (*Brassica chinensis*)	5.8	71	1.7	0.7	100	2.6	2.3	55	?
Edible Ferns (*Athyrium esculentum*)	8.5	109	3.4	0.8	25	4.4	2.3	10	?

[a] Source: Adapted from Oomen and Grubben 1978:36–37.

TABLE V

Energy and protein productivity of selected Pacific Island food crops[a]

	Energy production (Gigajoules/hectare)	Protein Production (kg/ha)
Cassava	34.3	37
Yam	23.8	107
Sweet potato	30.9	96
Maize	13.4	82
Rice	13.4	72

[a] Source: Adapted from Coursey and Booth 1977b:76.

basis) and the "possibility of using the leaves in combination with the roots to counterbalance the deficiency of protein in the latter (especially for livestock feed) has attracted some attention in recent years" (Coursey and Halliday 1977:141). Martin and Ruberte (1979:33) state that cassava can be considered one of the best of the tropical green leaf vegetables because of its high protein, vitamin and mineral content.

Cassava leaves are already a very importent food in West Africa, Indonesia, Malaysia and parts of Brazil, and in Nigeria special lines are cultivated which produce very few tubers but have a high production of palatable leaves free from mosaic virus symptoms (Oomen and Grubben 1978:101). While an important part of West African diets, leaves are eaten in the Pacific only on Fiji's Lau group, in Vanuatu (formerly the New Hebrides) (McGee, Ward and Drakakis-Smith 1980:186) and occasionally in Tonga (Thaman 1976:144).

While cassava is one of the most efficient producers of carbohydrates known (Coursey 1978:132) (Table V), the tuber is low in both quality and quantity of protein. The tuber flesh is composed of "about 62% water, 35% carbohydrate, 1–2% protein, 0.3% fat, 1–2% fiber and 1% mineral matter. . . . most of the carbohydrate is a starch" (Onwueme 1978:145). Table VI compares the composition of cassava with other common Pacific Island staples. The minerals in the tuber include calcium, phosphorous and iron. The tuber is relatively rich in Vitamin C (34 mg per 100 g fresh weight) and contains traces of niacin as well as traces of thiamin and riboflavin. Onwueme (1978:145) mentions that there are also traces of Vitamin A. Schofield (1979:59) supports this and says that although cassava diets are characteristically deficient in protein, they usually have surpluses in vitamins A and C. Furthermore, according to Coursey and Booth (1977b:81) "the protein content of cassava roots when compared with other food crops such as rice on the basis of weight of protein per 100 calories is not so unfavourable as is generally believed."

It should be stressed that both leaves and tubers of cassava contain small but significant amounts of very poisonous cyanogenic glucosides which have been known to cause death to consumers of cassava (Dharmawardene unpublished observations 1980).[2] These glucosides are "linamarin and lotaustralin . . . both of which are highly soluble in water and tend to decompose when heated to temperatures of about 150° C. Under the influence of the enzyme linamarase which is also found in cassava, both glucosides are hydrolysed to produce hydrocyanic acid (prussic acid) which is highly poisonous to humans and animals" (Onwueme 1978:145), but can be easily poured off with the water. Whereas the detoxification of the tubers of the bitter varieties of cassava is time consuming, the sweet varieties store most of the poisonous glucosides in the peel and can

TABLE VI

Composition per 100 g uncooked edible portion of selected Pacific staple crops[a]

SPECIES	DRY MATTER g	ENERGY MG	PROTEIN g	FIBER g	CALCIUM mg	PHOS-PHOROUS mg	IRON mg	CAROTENE mg	ASCORBIC ACID mg	THIAMIN mg	RIBO-FLAVIN mg	NIACIN mg
Cassava Tuber (raw)	34.5	564	1.0	1.0	26	32	0.9	0	34	.05	.04	0.6
Cassava (starch)	88.0	1480	0.5	—	0	0	0	0	6	0	0	0
Yam (*Dioscorea alata*)	23.6	364	1.9	0.6	38	28	1.1	.005	6	.10	.04	0.5
Sweet potato	27.7	451	1.0	0.8	21	50	0.9	.04	21	.14	.05	0.7
Colocasia taro	24.6	393	2.2	0.8	34	62	1.2	trace	8	.12	.04	1.0
Breadfruit	27.1	401	1.3	1.3	29	40	0.7	.01	12	.08	.06	1.2
Cooking banana	31.8	468	0.9	0.4	18	38	0.6	.48	11	.15	.06	0.7
Maize (whole kernel dried)	86.4	1459	9.1	2.3	14	245	2.8	.27	0	.29	.11	2.1
Rice (milled polished)	88.2	1530	6.4	0.3	24	135	1.9	0	0	.10	.05	2.1
Wheat Flour (white)	88.0	1522	8.6	0.2	17	350	1.0	0	0	.18	.05	1.3
Irish potato	21.7	343	2.0	0.4	9	52	0.8	trace	18	.10	.04	1.6

[a] Source: Leung, Butram and Chang 1972.

even be eaten raw by livestock. In many areas the uncooked peels are fed to domestic animals (Onwueme 1978:147).[3]

Despite the low and poor quality protein content of cassava and its high glucoside content, it has considerable potential in the direct production of protein when used as an animal feed. Coursey and Halliday (1977:140) report that raw cassava with hydrocyanic acid concentrations of 400 mg per kg has been fed experimentally to pigs as their sole carbohydrate source without apparent ill effects. Gerpacio *et al.* (1977a, 1977b) have also stressed the potential of high proportions of cassava in swine and broiler chicken rations, and Enriquez and Ross (1972) showed that as much as 50% cassava meal in rations could be fed to chickens without any detrimental effects on egg production, feed conversion, egg weights, shell thickness, body weight or mortality.

Most producing countries cultivate cassava for human consumption. Brazil is a notable exception where it is estimated that between 1964 and 1966 cassava as stock feed accounted for 47% of average annual livestock production (Phillips 1974 in Coursey and Halliday 1977:140). In the Pacific it is increasingly important as stock food and studies in Papua New Guinea show that pigs were able to subsist on a diet of nothing but raw sweet cassava tubers and leaves for long periods with no visible deleterious effects (Quartermain, personal communication, 1980)[4] In Samoa a compound pig feed has been produced using both dried tubers and leaves (Mauala 1977:168). In non-producing countries the use of cassava in livestock feeds is growing. In 1975 about two million tonnes of cassava, mostly from Thailand and Indonesia, were exported overseas, mainly to European Economic Community countries (Coursey and Booth 1977a:101).

INTRODUCTION, DISTRIBUTION AND IMPORTANCE

The cassava invasion of the Pacific Islands apparently began in Micronesia and has most recently reached western Melanesia (Appendix 1 and Fig. 1). It was introduced into Western Micronesia during the time of Spanish colonization (Barrau 1961:53), whereas its entry into Polynesia and Melanesia began after 1800 (Massal and Barrau 1956; Barrau 1958, 1961; Brookfield and Hart 1971). In the more isolated areas and larger islands of the Solomons and in highland Papua New Guinea, it is a more recent introduction (Eele 1978a:58 1978b; Bourke 1982:125).

In Papua New Guinea it is a dominant crop in the dry savannas and around urban areas; its cultivation is expanding in the highlands and in the island groups such as New Britain, New Ireland and Manus (Powell 1977; Bourke 1976; Thaman 1977b, 1977c). In the Solomons it is of increasing importance, but a major crop only in the Guadalcanal savanna areas, around Honiara and on some of the smaller high islands with high population densities such as the Polynesian outliers of Tikopia and Anuta (Larson 1966; Yen 1979; Eele 1978a).

Cassava is extensively cultivated in many areas of New Caledonia and Vanuatu, especially in the drier zones of New Caledonia (Barrau 1958; Guiart 1956); it is also the dominant crop in urban gardens in Noumea and in the alluvial soils of some areas on the windward east coast of New Caledonia. In both areas it has often displaced the much more labor demanding yam and taro cultivation (Bonnemaison 1974).

In Fiji and Tonga cassava is now the most widely cultivated root crop, on both a subsistence and commercial basis in Fiji, and on a subsistence basis in Tonga. In Fiji it is by far the most frequently consumed food amongst the Fijian population (Lucas 1978:30) and is also very commonly cultivated and eaten by the Indian population. It is the cheapest and most abundant crop at the Suva Municipal Market (Thaman 1976/1977), costing only 17 Fijian cents per edible portion compared to breadfruit, plantain, *Xanthosoma* taro, banana, potato (imported) and *Colocasia* taro which were 24, 26, 28, 31, 38 and 40 Fijian cents (1 Fijian Dollar = 0.97$U.S. at 1984 exchange rates) respectively. Cassava was purchasable at a cost of only 1.1 cents per 100 calories compared to breadfruit, plantain, *Xanthosoma* taro, banana, potato and *Colocasia* taro at 2, 1, 2.4, 2.7, 5.1 and 3.5 cents respectively (Baxter 1977:36).

In Fiji's Lau Islands to the east, where cassava was introduced as part of a hurricane relief program (Lessin and Lessin 1970:129), it is now a major staple. According to Thompson (1940 in Bedford 1976:1), in Kabara in Southern Lau, where land was poor and food scarce, cassava was introduced in 1936 to help people increase food crops. There was only limited subsistence gardening on the island before the introduction of manioc and sweet potato.

In Tonga cassava is definitely the most important and frequently eaten subsistence crop in the low-lying Ha'apai group and is very important in the northern Vava'u group. A 1971–72 random survey of 101 bush allotments showed that, on the main island of Tongatapu, cassava comprised 37% of all subsistence cultivation compared with around 25 and 21% for *Xanthosoma* taro and yams (Thaman 1975:156, 1976:236–237, 1977a). In the same survey, it was found tht 73 of 87 cultivated sample bush allotments had cassava, compared with 58 and 52 for *Xanthosoma* taro and yams (Thaman 1975:156). Cassava is also the cheapest staple root crop at the Talamahu Market in the Tongan capital city of Nuku'alofa (Ward and Hau'ofa 1979:380). However, with the decline in export banana production, expanding overseas markets for taro, high local price of yams and a rapid expansion in cooperative yam gardens (Thaman 1978), there may well be a current shift away from cassava back towards yams and *Colocasia* taro as dominant staples.

Western Samoa is one area where cassava is not cultivated extensively. What is grown is used almost exclusively as pig and chicken food or for making starch by grating the roots (Mauala 1977). In American Samoa it is often cultivated for food around the town of Pagopago (Malcolm 1954; Merrick 1977). On Niue cassava is grown on the poorer soils in the central basin (Wright and van Westerndorp 1965; Yuncker 1943; Sykes 1970); in 1976, 94 of 100 sample households had some cassava (Mitchell 1977:123–124). In the southern Cook Islands cassava is reported to be the most widely cultivated crop in terms of area (Johnston 1967; Michael 1977; Raoult and Jabre 1976). On the island of Aitutaki it is the main food and is prepared in a powdered form for export to Pacific Islanders living in New Zealand (Ŕaoult and Jabre 1976:7).

In the Society Islands of French Polynesia, where it was introduced in 1850 (Barrau 1970:13), and in the Marquesas and Wallis and Futuna cassava has also become a major crop. In the Marquesas it is a dominant staple and on the small island of Mai'ao off Tahiti it had replaced yams as the basic staple in the diet of the people by early this century (Finney 1973:57). Its primacy on Mai'ao, however, apparently has recently declined due to an overemphasis on commercial coconut production, but it remains a dominant staple crop in most of high islands French Polynesia.

Since its early introduction into the former American Trust Territory of Micronesia, cassava has become a prominent staple in almost all high island groups except possibly Ponape, where yams are still widely cultivated (Stone 1970). As a result of taro diseases, possibly *Phytophthora*, cassava became the staple crop during the German period in Yap, where there are now about 20 to 25 cultivars (Barrau 1961:53). High starch varieties were also cultivated on a commercial basis by the Japanese for export back to Japan during the Japanese occupation of Micronesia (Farrell 1972).

In the low-lying atolls of Tuvalu, Kiribati, the Northern Cook Islands and the Tuamotus, cassava is not cultivated and inhabitants depend on coconut, breadfruit, pandanus, some bananas and the giant swamp taro (*Cyrtosperma* spp.) as their local staples. In the late 1940s when Tuvaluans (Ellice Islanders) from the island of Vaitupu were resettled on the island of Kioa off Vanua Levu in the Fiji Islands and faced initial food shortages, cassava was generally the first food plant cultivated in new gardens. In the early 1960s it was still the principal staple, despite its being a non-traditional crop on the atoll soils of Vaitupu.

On Nauru, Kiribati and Tuvaluan indentured workers commonly cultivate cassava in planter boxes in soil transported to their tenement housing areas. Over 25% of 172 Kiribati and Tuvaluan gardens studied by Thaman in 1981 had cassava. White (1965) mentions that in Hawaii cassava is widely grown for its edible roots and is boiled and eaten like potatoes or grated and used as a substitute for bread or for starch and tapioca.

Cassava has become a major staple of increasing nutritional and economic importance in most Pacific Island groups. In fact we have even seen cassava, which was reportedly introduced into Australia by indentured plantation laborers from the Pacific Islands, cultivated for food today by Fijian and Tongan residents in urban Brisbane. It is currently being cultivated experimentally at Coolum Research Station in South Queensland and in Bathurst, New South Wales as a possible source of ethanol, in preference to sugar cane, as cassava can be harvested year round (Evans personal communication, 1979).[5]

IMPLICATIONS OF THE INVASION

Cultural Implications

Food in the Pacific was traditionally not merely a means of sustenance but had great social, political, religious and in some cases magical significance. This is becoming less true, especially in urban areas. There has been a trend toward the consumption of nutritionally-inferior foods which are easy to grow, prepare and store. Cassava has become the locally-produced "TV dinner of the Pacific" (Forge 1980: personal communication).[6]

With the almost simultaneous introduction of cash cropping and cassava into the Pacific, local food systems have been confronted with the forces of modernization. This has led to a deterioration of the heterogeneous and highly effective traditional agricultural and food systems, and a replacement of labor-intensive and environmentally-demanding crops such as taro and yams by less labor- and space-demanding crops like cassava. These same forces have also led to a simultaneously increasing dependence on imported foodstuffs such as polished rice and white flour.

While cassava cannot be implicated in the initial breakdown of social or food systems, it has certainly played a primary role in assisting gastronomic and cultural change by providing easily grown, cheap and filling food. Its very utility has provided the "ammunition" for the cassava invasion. Given rising prices for both marketed local crops and imported foods, its price remains relatively stable. Vila market is a good example. Over a six-year period from 1971–1976, cassava increased 33% in price while bananas and yam increased by more than 100% (McGee, Ward and Drakakis-Smith 1980:95).

As the more and better quality lands have been planted in cash and long-term tree crops, there has been insufficient quality land for traditional staples. As more time and effort have been required to cultivate cash crops, and as more young men have moved out of the rural areas to take up wage employment, there has been insufficient time or labor for the labor-intensive traditional crops. Population pressure and schooling, which has effectively removed the young labor force for most of the day, have aggravated the problem. There has been neither land nor time, and frequently insufficient motivation, to continue the more demanding cultivation of traditional staples. As pressure on land increased and fallow periods became shorter, and as soil fertility and yields of traditional staples decreased, some accommodation was obviously needed. Almost without exception the answer throughout the Third World has been cassava! What else has produced as much for so little? Were there any alternatives? Further dependency on expensive imported foods of almost equally low nutritional value could not be sustained by the insular economic systems.

Although there are Pacific societies such as Western Samoa, where increased cash cropping has not meant a large increase in cassava production for human consumption, the symbiotic relationship between cassava and modernization can be seen clearly throughout the Pacific. There are areas where cassava is of little importance and where there is only partial dependence on cash cropping, such as on the Weather Coast of Guadalcanal (Chapman and Pirie 1974; Teteha 1976) and in highland Papua New Guinea. Intermediate are Vanuatu and areas of the Solomon Islands and Papua New Guinea where the cassava invasion is particularly noticeable in villages near urban areas. Then come Fiji, Tonga and Aitutaki in the Cook Islands where there is considerable cash cropping and a heavy dependence on cassava, most particularly in drier areas. Finally, there is almost total dependency on both cash cropping and imported food. On the island of Mai'ao where all available time was spent producing copra, the once dominant staple cassava has been abandoned in favor of bread, rice and pilot crackers (Finney 1973:59). In Fijian villages in the vicinity of tourist hotels or urban centers where villagers are full-time employees and very rarely engage in any form of gardening, most food-stuffs are purchased (Thaman 1982). The most extreme example is Nauru where the indigenous Nauruans of this phosphate-rich equatorial island subsist almost entirely on imported foodstuffs.

In the Pacific, cassava is regarded as a particularly low status food (Parham 1972:136; Howlett 1967:54; Massal and Barrau 1956; Hau'ofa 1979:99), and traditional staple crops such as taro and yams are now often reserved for special occasions. Increasingly, canned foods, bread, beer, rice and tinned fish, and even cassava are replacing traditional staples as status food. Moreover, the integral place of food as a social, religious and political force in many Pacific societies has virtually vanished.

The daily diets of many Pacific Islanders now include cassava in some form, either baked or boiled as a vegetable, or as the starch component in the various coconut and banana flavored puddings and porridges. Cassava starch has almost completely replaced the traditional starch made from the Polynesian arrowroot (*Tacca leontopetaloides*), which was used in puddings and as an adhesive for tapa throughout Polynesia (Fa'anunu 1977; Thaman 1976). In 1971 survey results from Tonga indicated that cassava was eaten 340 days of the year (Thaman 1975:84). In Fiji Lucas (1978:30) states that cassava is the most frequently consumed food and in a survey of 321 meals, 40% of all lunches in the urban areas and 19% in the rural areas contained cassava. In Honiara *ghola* (baked, grated cassava) is the most common cooked food found in the market (Lasaqa 1966:58).

The introduction of cassava into the Pacific has obviously had vast implications on both food and social systems, but the changes might have been more dramatic and socially and nutritionally destructive had such an easily grown crop not been introduced. Lack of such a crop may have increased dependency on imported food or slowed down the spread of cash cropping or led to even greater malnutrition problems.

Nutritional Implications

Although little research has been carried out in the Pacific Islands on the nutritional impact of increased consumption, apparently malnutrition, most particularly in the form of kwashiorkor, can be attributed to diets predominant in cassava. Studies in Africa show that "peasant commur' 'es that rely heavily on cassava diets usually have a high incidence of protein malnutrition" (Onwueme 1978:157).

> "Mapping the disease has helped to make it clear that kwashiorkor mainly occurs in regions where children are brought up on a basic diet of cassava and bananas. The disease is rarely observed on the other hand, in areas with a diet of cereals such as millet. These differences can be explained by the varying protein of these plants" (Manshard 1974:136).

There are also considerable malnutrition problems in areas of Africa where children are weaned onto a thin gruel made from cassava (Lowenberg *et al* 1968:224). These examples of problems related to the low protein content of cassava underline the importance of eating it with other foods, as it has been estimated that 25 pounds (11.4 kg) of cassava made into tapioca must be consumed daily to meet an individual's protein requirements (Anderson 1972:122).

There are other problems associated with diets predominant in cassava. It has been suggested that "conditions such as goiter may result from long-term habitual consumption" (Martin and Ruberté 1975:29). In central Africa it has been established that goiter and cretinism are caused by the joint effects of iodine deficiency and a goiterogen found in cassava (Ermans *et al* 1980). This goiterogenic effect may have important implications for some highland and non-coastal groups in Papua New Guinea and Fiji where endemic goiter is a problem.

In the Pacific nutritional levels, particularly of children, have deteriorated over the last 10 years. There are increasing cases of PEM (protein energy malnutrition), kwashiorkor, goiter, iron-deficiency anemia and beriberi (Lambert 1979; Parkinson 1977; Mackenzie 1976/77; Thaman 1979). The role of cassava in this trend is currently not known. With respect to increasing PEM, preliminary studies in Papua New Guinea indicate that insufficient calories rather than protein deficiency may be the limiting nutritional factor responsible for increasing malnutrition (Heywood 1982). Thus cassava, with its remarkable production of calories, could have a major future role to play in minimizing PEM in the Pacific Islands. Changing traditional breastfeeding and childrearing practices and an increasing dependency on imported foodstuffs with corresponding declines in fresh foods and vegetables, particularly in urban areas, may be held primarily responsible for the decline in nutritional levels in the Pacific.

Another advantage seen of cassava seems to be its relatively high fiber content in contrast to the majority of imported or processed foods. While the actual implications of fiber content in the diet are not fully understood, there is evidence that foods containing a relatively high fiber content assist in combatting diseases of the colon, diverticulitis, heart disease, diabetes, cancer and ulcers, all of which are increasingly common in the Pacific (Tudge 1977:79–81).

A further nutritional benefit is the increasing use of cassava as animal feed. This could increase available protein and also free more nutritious foods (such as coconut, papaya and sweet potato- commonly fed to chickens, pigs and dogs) for human consumption. Where cassava has not become an accepted staple food crop, as in parts of the Papua New Guinea Highlands, on the Weather Coast of Guadalcanal in the Solomon Islands and in Western Samoa, cassava is cultivated almost exclusively as pig and chicken feed.

Environmental Implications

While cassava is widely held responsible for soil degradation and nutrient depletion (Massal and Barrau 1956), as we have mentioned earlier, this does not appear to be the case. It is true, however, if cassava is repeatedly cropped on already depleted soils, as it frequently is, usually with no addition of fertilizer or mulching, further soil degradation can be expected, but at a lesser rate than caused by many other tropical crops. Moreover, cassava may also have a beneficial effect in extracting and recycling nutrients on improverished soils.

Because if its broad environmental tolerance, cassava is planted in many parts of the Pacific in poor soils on steep slopes that are unsuitable for any other crop. When planted in these conditions, particularly as a single crop on completely cleared slopes or grass-lands, accelerated soil erosion can occur, especially in the early stages of growth before the leaf canopy has closed. This is particularly serious when planted before or at the onset of the rainy season. As a crop characterized by a complete canopy, little grows around the roots or under the plants, so when the crop is harvested, the land is left open to increased weathering and leaching. Furthermore, the common practices of mounding and ridging when planting cassava in savanna areas seem to accelerate erosion as it loosens the soil (Nye and Greenland 1960:89–90). Nye and Greenland (1960:91) go on to stress that "repeated cycles of cropping and fallowing may however cause total ruin to savanna soils because they uncover subsoils rich in iron oxide, which if not indurated already, become so on exposure."

Related to the problems of vegetation removal and continuous cropping are the possible detrimental changes in runoff, water table levels and the overall hydrologic regimes in areas under intensive cropping.

Finally, when cassava is planted as a single crop three or more times in succession, the intensification of the cropping cycle results in a very significant decline in both the total biomass of the area and in floral and faunal diversity (Clarke 1966; Manner 1976, 1977). Although not directly responsible for such changes, cassava has played and may continue to play a major role in the improverishment of Pacific Island ecosystem diverstiy and stability.

CASSAVA AS AN ADAPTATION TO CHANGE

Processes of modernization such as increased cash-cropping and monetization supported by export-oriented and industrial economic policies, institutionalization of education, communication and work, and associated shifts of population to urban areas are contributing to major changes in Pacific Island economic and social life. Five major areas are examined in relation to Pacific food systems: decreasing labor availability for domestic food production; decreasing land availability for domestic food production; food scarcity and increasing food dependency; environmental deterioration; and general social and cultural modification. In most cases cassava can be seen as being almost synonomous with change, as it is either partly responsible for or its adaption is in response to, such changes.

Labor Scarcity

In many Pacific villages there is an increasing shortage of labor for domestic food production. The educational system occupies the young labor force in the classroom rather than in the garden. Interest and availability of recreational activities such as sporting events and visits to town and other villages also reduce time spent in the fields. Labor mobility, particularly among young adults, results in a diminished rural labor force and more wage demands in urban areas.

Government promotion of export and non-food crops has also affected worker availability for production of traditional labor-intensive food crops and has meant that women are increasingly responsible for food production. Furthermore, with social and government pressures toward modernization and the need for cash for a variety of daily social and economic obligations, villagers have had little option but to change to cash cropping.

Adoption of cassava as the staple food crop has allowed villagers to participate in the cash economy and at the same time to meet the problem of labor shortage for local food production. Ability to address such circumstances helps preserve the people's "food independency," a vital consideration when world prices for food, transport and fertilizers are increasing. In the urban areas, particularly squatter areas or in work camps where there is considerable population mobility, cassava, because of its low labor and cost input, is commonly planted, even if there is uncertainty whether the planter will in fact remain long enough to harvest the crop. In these areas it is often the only locally grown food available. Cassava may provide a useful "transition" food system between the totally traditional self-sufficient societies where labor was abundant and the fully monetized society where labor for food production is scarce.

Land Scarcity

In many Pacific rural areas there is no longer sufficient land for traditional food crops. High quality land near villages has been pre-empted for cash and export crops. Long-term tree crops such as cocoa, coffee and rubber tie up land for long periods, and government promoted cattle schemes tie up large areas of land that were once part of the food crop and fallow systems. Coffee in Papua New Guinea, cocoa in Western Samoa and Fiji, cattle projects in Papua New

Guinea, the Solomon Islands and Fiji, and pineapple farming in the Cook Islands are examples of this. Moreover, many traditional staples such as yams and taro require fertile soils which necessitates long fallows. Pre-empting good land for cash cropping has, therefore, led to heavy pressure on other less fertile village lands and a shortening of the fallow period. As a result much peripheral village land has become severely degraded and will no longer support traditional staples. However, it will support cassava. This is seen by local villagers as preferable to cultivating land at greater distances from the village which requires energy and time in getting to the gardens and precludes the possibility of policing them against theft and animal damage.

In many Fijian villages and in the Tongan islands of Ha'apai, cassava provides an adaptation to a land shortage caused by increased export cropping and population pressure. Similarly, in urban or peri-urban areas, cassava is found growing on vacant lots, land awaiting development, road frontages and slopes too steep for buildings. Cassava seems to counter the scarcity of urban land available for food production.

Food Scarcity and Food Dependence

With rapid worldwide inflation and incommensurate wage increases, nutritious, imported and local foodstuffs are very costly. For example, in Tonga where the average daily wage is between $3 and $5, a basket of yams (family carbohydrates for two to three days) may cost as much as $15 to $20, a string of fish $5, or a can of tinned fish $1. Cheaper, nutritionally inferior foods such as polished rice, bleached flour, sugar or cassava become the only alternatives for those who must purchase their food. As labor demanding crops such as taro and yam are less frequently planted, their prices climb even higher. The only alternative to almost complete dependence on imported food is cassava. It is cheap and a good source of carbohydrates, fiber and some vitamins. Most important, it is local! Local production of cassava also frees money which would have been spent on other staples for the purchase of other nutritional food to supplement a cassava diet.

Food shortfalls in areas such as Yap in Micronesia, Manus in Papua New Guinea or in the Solomon Islands, where the traditional staple taro can no longer be cultivated due to taro blight (*Phytophthora colocasiae*), will most probably by filled by either sweet potatoes, cassava or imported rice and wheat flour. The role of cassava as the "stable staple" may become even greater due to recent major outbreaks of disease affecting three other Pacific Island staple crops: sweet potatoes, yams and bananas.

Finally, another aspect of food supply shortages throughout the Pacific concerns institutional feeding in boarding schools, military barracks, hospitals and the like where large numbers must be fed on limited budgets. The answer again is usually rice or cassava.

Cassava is the most common and possibly most successful local answer to the increase in imported food prices and the shortage of cheap traditional nutritious food. It provides a local staple that is little affected by inflation and unemployment. For example, in New Caledonia, when world nickel prices fell, workers in tenement housing in Noumea began planting cassava on undeveloped land. Similarly, when funds are short, students at boarding schools and at the University of the South Pacific in Fiji get cassava instead of the more preferred taro.

Although apparently of considerable benefit in helping Pacific Islanders adapt to the processes of modernization, cassava has the disadvantages of relatively low nutritional value and the presence of cyanogenic glucosides and goiterogens. Fortunately, glucoside concentrations are low in the sweet varieties grown in the Pacific.

Environmental Deterioration

Modernization has led to considerable environmental degradation in the Pacific. There is decreasing open space, water shortage and land scarcity. Environmental deterioration also includes accelerated soil erosion, soil nutrient depletion and increasing use of and need for costly and dangerous pesticides.

Cassava cultivation and consumption has enabled Pacific peoples to adapt to these changing environmental conditions. It is a crop that yields food from eroded, nutrient-deficient soils. It can be planted despite water shortages and is pest and disease resistant, obviating the use of pesticides. Where water rationing is a common occurrence as in Port Moresby and in Western Viti Levu, Fiji, cassava is one crop that survives when water use for food gardening is prohibited.

Social and Cultural Modifications

Aspects of cultural modification to be considered are changing food systems; decreasing reciprocity and ceremonial exchange; changing division of labor and breakdown in the communal or family work ethic.

Change within traditional food systems through replacement of crops such as yam and taro by cassava are not as socially disruptive as they initially appear. Cassava is a root crop with cultivation practices similar to traditional staples. It can be grown, cooked and marketed in much the same way.

Systems of reciprocity and ceremonial exchanges have been modified. Increasing monetization and the decline in production of the traditional foods and crafts used in ceremonial exchange has led to their limited availability. Increasingly, Pacific Islanders have to purchase these items or replace them with imported substitutes. As purchased items are very expensive, this has led to a significant limitation of ritual exchange.

Although the substitution of cassava for traditional foods in these ceremonies has changed their nature, it has provided a food that is within most islanders' financial means. In Tonga and Western Samoa, for example, cassava would never be presented at an important wedding or feast, but it is now the common fare for feeding people at funerals in Tonga and is now acceptable at Fijian feasts. Thus cassava does have a role in ritual exchange, both through direct exchange and through being served at feasts.

Social changes seen in the division of labor and communal or family work ethic seem to by directly related to the increased involvement of men in cash cropping or cash employment, the involvement of young labor in education and urban life styles, and general breakdown of the extended family as the traditional production unit.

With the men and school children devoting less time to food production, women have taken on this role, especially in places such as the Solomon Islands, Western Samoa and Fiji. Cassava, easily planted and maintained, possibly minimizes the effects of such social changes. Men and school children can often provide enough food for the family by working sporadically after work or on the weekends, thus helping women in food production and allowing them to devote more time to other efforts. The very fact that large amounts of food can be produced provides a continuing base for supporting the larger extended family.

SUMMARY

In summary, although cassava has played a role in cultural and environmental degradation, it has also permitted Pacific peoples to adapt to urbanization, modernization, cash cropping and cash employment by allowing them to eat a local familiar starchy staple. It also allows them to garden as they did before and provide at least some measure of escape from the psychological, sociological and environmental troubles of a changing world.

Cassava is still considered an inferior food in many areas of the Pacific Islands and is of relatively low nutrient status when compared with most traditional Pacific Islands staple food crops. However, it has become a major staple in many Pacific Island countries, and it is a prodigious producer of valuable calories at a time when there is increasing protein energy malnutrition (PEM). Cassava can be grown on the poorest and most degraded soils in both wet and dry areas; it is a common livestock feed; and it has edible leaves rich in nutrients. Cassava also is a possible flour supplement for breadmaking, and at a time of impending energy crisis, is a very promising renewable source of the fossil fuel alternative ethanol.

Cassava has sufficient favorable characteristics to warrant its serious consideration for widespread cultural, nutritional and ecological development in Pacific Island food systems. We present eight suggestions to take advantage of the integration of cassava into existing island agroecosystems (Janzen 1973) and to maximize cassava's role in the food and agricultural future of the Pacific Islands. Our development suggestions are:

That education and media campaigns be initiated to emphasize the nutritional value of both the tubers (as an inexpensive food energy source) and the leaves (additional nutrients); proper ways of preparation to avoid possible toxic effects and maximize nutritional value and palatability; and the value of cassava as a major "national nutritional resource" when eaten with other locally available foods.

That efforts be made to collect, classify and conduct field trials with local germ plasm and cultivars, as these are probably particularly suitable to local environments. Cassava clones in the Pacific are relatively disease-free, and there is, according to Nestel (1977), considerable risk in introducing new cultivars because of problems with mosaic virus in Africa and bacterial blight in Malaysia.

That local research on intercropping cassava with both commercial and other subsistence crops be conducted in an attempt to diversify both commercial and subsistence cassava plantings.

That suitable rotation systems with cassava be developed, possibly with periods of improved fallow or grass leys[7] in an attempt to minimize the possible development of indurated hardpans under intensive successive plantings.

That other soil conservation techniques be encouraged in cassava growing areas to include planting trees; strip cropping of cover crops along the contours; prohibiting the planting of ridge tops; encouragement of terracing (which is traditional practice in most Pacific Island groups) or the use of mulching and other means of organic fertilization which have shown to improve yields in other tropical areas (Nye and Greenland 1960:85).

That wider use of both the dried tubers and leaves be made for feeding livestock, both in commercial rations which could be substituted for costly imported rations, and as a substitute for other local subsistence foods which could be diverted to human consumption. Allen *et al.* (1978:5) have suggested as a possible means of improving child nutrition in the Nembi Plateau in Papua New Guinea's Southern Highlands, "the use of cassava as pig food, releasing more sweet potato production for human consumption."

That the importance of cassava as a staple food crop on undeveloped lands in densely settled urban areas be assessed, as well as the potential of cultivating individual plants in backyard gardens or in hedges as sources of nutritious leafy greens.

That fruit trees, firewood, timber species and other culturally valuable tree species be protected when clearing areas for cassava cultivation and, where possible, establish vegetation reserves in individual or clan garden areas which might serve as gene pools of culturally and ecologically valuable plant species.

The development of diverse agroecosystems in which cassava provides a major source of food energy is a practical challenge facing the Pacific Islands. Preservation of the agricultural, nutritional and ecological diversity characteristic of most traditional Pacific Island food systems must also be addressed. Given a rational approach to its use, cassava does offer an important alternative to increasing dependency on foreign foods in the Pacific. Socially and economically sustainable self-sufficiency in foodstuffs helps both the individuals and the nations of the Pacific. With different consumption patterns (including both leaves and tubers), cassava could provide the needed nutritional, economic and cultural security against unforeseen and often uncontrollable economic crises and natural disasters which can lead to increasing food dependency in the Pacific. Therefore, a basic challenge facing Pacific planners and decision-makers is to effectively integrate modernization and cassava in the hope of minimizing further deterioration of Pacific Island food systems and their associated cultures and environments. As stressed by Martin (1970), if used wisely, there *is* a very important role for cassava in the "world of tomorrow."

NOTES

1. Cassava was observed by K. Schwerin in 1970 growing at 2000 m (6560 ft.) near Girón in southern Ecuador (lat. 3° 10′ S) (Editor).
2. from an unpublished paper, "Cassava: The Food and Poison," presented at the School of Natural Resources, University of the South Pacific, Suva, Fiji in 1980.
3. Carl Spath, in an unpublished paper, cites Clark to the effect that large scale consumption of manioc derivatives may produce fatty degeneration of the liver and kidneys. Osuntokun concludes that heavy reliance on manioc in the diet of poor Nigerians is the principal cause of tropical ataxic neuropathy. This disease is characterized by irreversible nerve degeneration, "predominantly posterior column myelopathy ... closely followed by optic atrophy and perceptive deafness" (Osuntokun 1973:133). In advanced stages the disease also involves demyelination of the peripheral nerve sheaths. In both cases the pathological condition is correlated with the physiological problem of detoxifying dietary cyanide (Editor).
4. Alan Quartermain, livestock husbandry expert, Papua New Guinea University of Technology, Lae, Papua New Guinea
5. John Evans, Coolum Research Station, Queensland Department of Primary Industries, Coolum Beach, Australia
6. Anthony Forge, Professor of Anthropology, Australian National University, Canberra, Australia
7. A grass ley is a fallow planted or kept in grass which is used as a fallow in an attempt to protect the area from soil erosion, to gain productivity, etc.

REFERENCES

Allen, B. J., R. M. Bourke, L. J. Clarke, B. Coghill, C. F. Pain and A. W. Wood. 1978. *A Preliminary Report on a Child Nutrition and Survey of the Nembi Plateau, Southern Highlands Province*. Port Moresby: Office of Environment and Conservation
Anderson, Gerald C. 1972. The head has a stomach. pp. 122–126 in *The Ecology of Man: An Ecosystem Approach*, ed. by Robert L. Smith. London: Harper and Row
Barrau, Jacques. 1950. *Preliminary List of Economic Plants in New Caledonia*. Technical Paper No. 6. Noumea, New Caledonia: South Pacific Commission
_____ 1958. *Subsistence Agriculture in Melanesia*. Bulletin No. 219. Honolulu: B. P. Bishop Museum
_____ 1961. *Subsistence Agriculture in Polynesia and Melanesia*. Bulletin No. 223. Honolulu: B. P. Bishop Museum
_____ 1970. *Useful Plants of Tahiti*. Dossier 8. Paris: Société des Oceanistes

Baxter, Michael W. P. 1977. *Food in Fiji: Aspects of the Produce and Processed Foods Distribution Systems.* Centre for Applied Studies in Development. Suva, Fiji: University of the South Pacific

Bedford, Richard D. 1976. *Kabara in the 1970s.* UNESCO/UNFPA Project on Population and Environment in the Eastern Islands of Fiji. Project Working Paper No. 3. Development Studies Centre. Canberra: Australian National University

Bonnemaison, Joel. 1974. *Espaces et paysages avaires dans le nord des îles d'Aoba et de Maewo.* Societé des Oceanistes. Paris: Musée de l'Homme

Bourke, R. M. 1976. Food crop farming systems used on the Gazelle Peninsula of New Britain. pp. 45–52 in *1975 Papua New Guinea Foods Crops Conference Proceedings*, ed. by K. Wilson and R. M. Bourke. Port Moresby: Department of Primary Industries.

_____ 1982. Root crops in Papua New Guinea. pp. 121–133 in *Proceedings of the 5th International Symposium on Tropical Root Crops, Manila 17–21 September, 1979.* Los Banos: Philippines Council for Agriculture and Resource Research

Brookfield, H. C. and Doreen Hart. 1971. *Melanesia: A Geographical Interpretation of an Island World.* London: Methuen

Chapman, Murray and Peter Pirie. 1974. *Tasi Mauri: A Report on Population and Resources of the Guadalcanal Weather Coast.* East-West Population Institute. Honolulu: East-West Center/University of Hawaii

Clarke, William C. 1966. From extensive to intensive shifting cultivation: A succession from New Guinea. *Ethnology* 5:347–359

_____ 1971. *Place and People.* Berkeley: University of California Press

Coursey, David G. 1973. Cassava as food: Toxicity and technology. pp. 27–36 in *Chronic Cassava Toxicity: Proceedings of an Inter-disciplinary Workshop*, ed. by Barry Nestel and Reginald MacIntyre. Ottawa: International Development Research Centre

_____ 1978. Some ideological considerations relating to tropical root crop production. pp. 131–141 in *The Adaptation of Traditional Agriculture: Socioeconomic Problems of Urbanization*, ed. by E. K. Fisk. Development Studies Centre, Monograph No. 11. Canberra: Australian National University

Coursey, David G. and R. H. Booth. 1977a. Contributions of post-harvest bio-technology to trade in tropical root crops. pp. 100–105 in *Regional Meeting on the Production of Root Crops (24–29 November 1975, Suva, Fiji): Collected Papers*, ed. by Michel Lambert. Technical Paper No. 174. Noumea, New Caledonia: South Pacific Commission

_____ 1977b. Root and tuber crops. pp. 75–96 in *Food Crops of the Lowland Tropics*, ed, by C. L. A. Leakey and J. B. Wills. Oxford: Oxford University Press

Coursey, David G. and D. Halliday. 1977. Developments in the use of tropical root crops as animal feed. pp. 139–143 in *Regional Meeting on the Production of Root Crops (24–29 November 1975, Suva, Fiji): Collected Papers*, ed. by Michel Lambert. Technical Paper No. 174. Noumea, New Caledonia: South Pacific Commission

Decker, Bryce G. 1970. *Plants, Man and Landscape in Marquesan Valleys, French Polynesia.* Ph.D. dissertation in Geography, University of California, Berkeley

Eele, G. J. 1978a. *1974–1975 Agricultural Statistics Survey: A Sample Survey of Solomon Islander Smallholder Agriculture.* Statistics Office. Honiara, Solomon Islands: Ministry of Finance; Ministry of Agriculture and Lands

_____ 1978b. Indigenous agriculture in the Solomon Islands. pp. 46–71 in *The Adaptation of Traditional Agriculture: Socioeconomic Problems of Urbanization*, ed. by Ernest K. Fisk. Development Studies Centre, Monograph No. 11, Canberra: Australian National University

Enriques, F. Q. and E. Ross. 1972. Cassava root meal in grower and layer diets. *Poultry Science* 51:228

Ermans, A. M., N. M. Mbulamoko, F. Delange and R. Ahluwalia. 1980. *Role of*

Cassava in the Etiology of Endemic Goitre and Cretinism. Ottawa: International Development Research Centre

Fa'anunu, H. O. 1977. Traditional aspects of root crop production in the Kingdom of Tonga. pp. 191-199 in *Regional Meeting on the Production of Root Crops (24-29 November 1975, Suva, Fiji): Collected Papers*, ed, by Michel Lambert. Technical Paper No. 174. Noumea. New Caledonia: South Pacific Commission

F. A. O. 1979. *1978 FAO Production Yearbook*, Vol. 32. Rome: Food and Agriculture Organization

Farrell, B. H. 1972. The alien and the land of oceania. pp. 34-73 in *Man in the Pacific Islands*, ed, by Ralph G. Ward. Melbourne: Oxford University Press

Finney, Ben R. 1973. *Polynesian Peasants and Proletarians*. Cambridge, Mass.: Schenkman

Fischer, John L. and Ann M. Fischer. 1957. *The Eastern Carolines*. Pacific Science Board, National Research Council. New Haven, Conn. National Academy of Sciences

Gallasch, H. 1976. Integration of cash and food cropping in the lowlands of Papua New Guinea. pp. 101-115 in *1975 Papua New Guinea Food Crops*, ed. by K. Wilson and R. M. Bourke. Port Moresby: Department of Primary Industries

Gerpacio, A. L., L. S. Castillo, C. C. Custodie, D. B. Roxas, N. M. Uichanco, V. G. Aranzosa and F. C. Arganzosa. 1977a. Cassava meal in swine rations. pp. 486-490 in *Regional Meeting on the Production of Root Crops (24-29 November 1975, Suva, Fiji): Collected Papers*, ed. by Michel Lambert. Technical Paper No. 174. Noumea, New Caledonia: South Pacific Commission

Gerpacio, A. L., D. E. Roxas, N. M. Uishanco, N. P. Roxas, C. C. Custodio, C. Mercado, L. A. Gloria and L. S. Castillo. 1977b. Tuber meals as carbohydrate sources in broiler rations. pp. 183-185 in *Regional Meeting on the Production of Root Crops (24-29 November 1975, Suva, Fiji): Collected Papers*, ed. by Michel Lambert. Technical Paper No. 174. Noumea, New Caledonia: South Pacific Commission

Guiart, Jean. 1956. *L'agriculture Vivriere Autochtone de la Nouvelle-Caledonie*. Noumea, New Caledonia: Commission du Pacifique Sud

Hansel, J. R. F. and J. R. D. Wall. 1976. *Land Resources of the Solomon Islands. Vol. 1: Introduction and Recommendations*. Land Resources Study 18. Surrey, England: Land Resources Division

Hardaker, J. B. 1971. *Kingdom of Tonga: Report on the Economics of Agriculture*. Armidale: University of New England

Hau'ofa, Epeli. 1979. *Corned Beef and Tapioca: A Report on the Food Distribution Systems in Tonga*. Development Studies Centre Monograph No. 19. Canberra: Australian National University

Haynes, P. H. 1977. Root crops production and use in Fiji. pp. 44-50 in *Regional Meeting on the Production of Root Crops (24-29 November 1975, Suva, Fiji): Collected Papers*, ed. by Michel Lambert. Technical Paper No. 174. Noumea, New Caledonia: South Pacific Commission

Haynes, P. H. and P. Sivan. 1977. Towards the development of integrated root crop research and production in Fiji. pp. 51-61 in *Regional Meeting on the Production of Root Crops (24-29 November 1975. Suva, Fiji): Collected Papers*, ed. by Michel Lambert. Technical Paper No. 174. Noumea, New Caledonia: South Pacific Commission

Heywood, Peter. 1982. Nutrition in Papua New Guinea and some implications for food production. in *Proceedings of the Second Papua New Guinea Food Crops Conference, Goroka, 14-18 July 1980*, ed. by R. M. Bourke and V. Kesavan. Port Moresby: Department of Primary Industries

Howlett, Diana. 1967. *Papua New Guinea: Geography and Change*. Melbourne. Thomas Nelson

Janzen, Daniel H. 1973. Tropical agroecosystems. *Science* 82:212-219

Johnston, K. M. 1967. *Village Agriculture in Aitutaki, Cook Islands*. Department of Geography, Pacific Viewpoint Monograph No. 1. Wellington, New Zealand: Victoria University of Wellington

Lambert, J. N. 1979. *Population Growth, Nutrition and Food Supplies. Population of Papua New Guinea*. ESCAP Country Monograph. New York: United Nations Economic and Social Council

Larson, Eric H. 1966. *Nukufero: A Tikopian Colony in the Russell Islands*. Department of Anthropology. Eugene, Ore.: University of Oregon

Lasaqa, I. Q. 1969. Honiara market and the supplies from Tasimboko West. pp. 48–96 in *Pacific Market Places*, ed.by H. C. Brookfield. Canberra. Australian National University

Lessin, Alexander P. and Phyllis J. Lessin. 1970. *Village of the Conquerors-Sawana: A Tongan Village in Lau*. Department of Anthropology. Eugene, Ore.: University of Oregon

Lowenberg, Miriam E., E. N. Todhunter, E. D. Wilson, M. C. Feeney and J. R. Savage. 1968. *Food and Man*. New York: Wiley

Lucas, V. E. 1978. *Malnutrition in Pre-School Children in Fiji*. New York: Foundation for the Peoples of the South Pacific

Lynam, J. K. 1978. Options for Latin American countries in the development of integrated cassava production programs. pp. 213–256 in *The Adaptation of Traditional Agriculture: Socioeconomic Problems of Urbanization*, ed. by Ernest K. Fisk. Development Studies Centre Monograph No. 11. Canberra: Australian National University

McGee, T. G., Ralph G. Ward and D. W. Drakakis-Smith. 1980. *Food Distribution in the New Hebrides*. Development Studies Centre, Monograph No. 25. Canberra: Australian National University

MacKenzie, Margaret. 1976/77. Who is a good mother? *Ethnomedicine* 4(½):7–22

Malcolm, Shiela. 1954. *Diet and Nutrition in American Samoa*. Technical Paper No. 63, Noumea, New Caledonia: South Pacific Commission

Manner, H. I. 1976. *The Effects of Shifting Cultivation and Fire on Vegetation and Soil in the Montane Tropics of New Guinea*. Ph.D. dissertation in Geography, University of Hawaii, Honolulu.

——— 1977. Biomass: Its determination and implications in tropical agro-ecosystems: An example from montane New Guinea. pp. 215–242 in *Subsistence and Survival: Rural Ecology in the Pacific*, ed. by Timothy P. Bayliss-Smith and Richard G. Feachem. London: Academic Press

——— 1980. *Buma Subsistence Agriculture: Patterns Processed and Problems on Malaita, Solomon Islands*. Suva, Fiji: University of the South Pacific

Manshard, Walter. 1974. *Tropical Apriculture: A Geographical Introduction and Appraisal*. London: Longmans

Martin, Franklin W. 1970. Cassava in the world of tomorrow. pp. 53–58 in *Tropical Root and Tuber Crops Tomorrow, Vol. 1: Proceedings of the Second International Symposium on Tropical Root Crops*, ed. by Donald L. Plucknett. Honolulu: University of Hawaii

Martin, Franklin W. and Ruth M. Ruberté. 1979. *Edible Leaves of the Tropics*. 2nd edition. Mayagüez, P. R.: Mayagüez Institute of Tropical Agriculture

Massal, Emile and Jacques Barrau. 1956. *Food Plants of the South Sea Islands*. Technical Paper No. 96. Noumea, New Caledonia: South Pacific Commission

Mauala, Nusi. 1977. Root crops production in Western Samoa. pp. 167–170 in *Regional Meeting on the Production of Root Crops (24–29 November 1975, Suva, Fiji): Collected Papers*, ed. by Michel Lambert. Technical Paper No. 174. Noumea, New Caledonia: South Pacific Commission

Maude, Alaric. 1965. *Population, Land and Livelihood in Tonga*. Ph. D. dissertation, Department of Geography, Australian National University, Canberra

_____ 1973. Land shortage and population pressure in Tonga. pp. 163–185 in *The Pacific in Transition*, ed. by H. C. Brookfield. London: Edward Arnold

Merrick, J. E. 1977. Status of root crops in American Samoa. pp. 122–124 in *Regional Meeting on the Production of Root Crops (24–29 November 1975, Suva, Fiji): Collected Papers*, ed. by Michel Lambert. Technical Paper No. 174. Noumea, New Caledonia: South Pacific Commission

Michael, G. 1977. Root crops production in the Cook Islands. pp. 92–98 in *Regional Meeting on the Production of Root Crops (24–29 November 1975, Suva, Fiji): Collected Papers* ed. by Michel Lambert. Technical Paper No. 174. Noumea, New Caledonia: South Pacific Commission

Mitchell, G. D. 1977. *Village Agriculture in Niue*. Publlished M. A. thesis in Geography, University of Canterbury, Christchurch, New Zealand

Neal, Mary C. 1965. *In Gardens of Hawaii*. Special Publication 50. Honolulu: B. P. Bishop Museum

Nestel, Barry L. 1977. Cassava—some recent research findings, pp. 1–5 in *Regional Meeting on the Production of Root Crops (24–29 November 1975, Suva, Fiji): Collected Papers*, ed. by Michel Lambert. Technical Paper No. 174. Noumea, New Caledonia: South Pacific Commission

Norse, David. 1979. Natural resources, development strategies and the world food problem. pp. 12–51 in *Food, Climate and Man* ed.by Margaret R. Biswas and Asit K. Biswas. New York: Wiley

Nye, Peter H. and D. J. Greenland. 1960. *The Soil Under Shifting Cultivation*. Commonwealth Bureau of Soils, Technical Communication No. 51. London: Commonwealth Agricultural Bureaux

Onwueme, I. C. 1978. *The Tropical Tuber Crops*. New York: Wiley

Oomen, H. A. P. C. and G. J. H. Grubben. 1978. *Tropical Leaf Vegetables in Human Nutrition*. Department of Agricultural Research, Communication 69. Amsterdam: Koninklijk Instituut voor de Tropen

Osuntokun, B. O. 1973. Ataxic neuropathy associated with high cassava diets in West Africa. pp. 127–138 in *Chronic Cassava Toxicity: Proceedings of an Interdisciplinary Workshop*, ed. by Barry Nestel and Reginald MacIntyre. Ottawa: International Development Research Centre

Parham, B. E. V. 1972. *Plants of Samoa*. Information Series No. 85. Wellington, New Zealand: Department of Scientific and Industrial Research

Parkinson, S. V. 1977. *The South Pacific Handbook of Nutrition*. Suva, Fiji: Y.W.C.A.

Phillips, Truman P. 1974. *Cassava Utilization and Potential Markets*. Ottawa: International Development Research Centre

Pimentel, David. 1979. Energy in agriculture. pp. 73–106 in *Food, Climate and Man*, ed. by Margaret R. Biswas and Asit K. Biswas. New York: Wiley

Powell, J. M. 1977. Plants, man and environment in the island of New Guinea. pp. 11–20 in *The Melanesian Environment*, ed. by John H. Winslow, Canberra: Australian National University

Purseglove, John W. 1968. *Tropical Crops: Dicotyledons*. 2 vols. London: Longmans

Raoult, André and Bushra Jabre. 1976. *Nutritional Problems in the Pilot Area of Aitutaki (Cook Islands)*. Noumea, New Caledonia: South Pacific Commission

Schofield, Sue. 1979. *Development and the Problems of Village Nutrition*. Sussex: Institute of Development Studies

Sevele, F. V. 1973. *Regional Inequalities in Socio-economic Development in Tonga: A Preliminary Study*. Ph.D. dissertation, University of Canterbury, Christchurch, New Zealand

Stone, B. C. 1970. The flora of Guam. *Micronesia* 6:1–65

Sykes, William R. 1970. *Contributions to the Flora of Niue*. Bulletin 200 Wellington, New Zealand: Department of Scientific and Industrial Research

222   R. R. THAMAN AND P. M. THOMAS

Templeton, J. K. and Michel H. Lambert. 1977. Proposals for a root crop development project in the Pacific. pp. 209–213 in *Regional Meeting on the Production of Root Crops (24–29 November 1975, Suva, Fiji): Collected Papers*, ed. by Michel Lambert. Technical Paper No. 174. Noumea, New Caledonia: South Pacific Commission

Teteha, Samson. 1976. *Land Use in the Tina River Basin. Weather Coast, Guadalcanal.* SE300 Research Paper in Geography. Suva, Fiji: University of the South Pacific

Thaman, Randolph R. 1975. Tongan agricultural land use: A study of plant resources. pp. 153–160 in *Proceedings of the International Geographical Union Regional Conference and Eighth New Zealand Geography Conference, Palmerston North, December 1974*, ed. by William Brockie, Richard LeHeron and Evelyn Stokes. Conference Series No. 8. Christchurch: New Zealand Geographical Society

———— 1976. *The Tongan Agricultural System: With Special Emphasis on Plant Assemblages.* Suva, Fiji: University of the South Pacific (published version of Ph.D. dissertation, U.C.L.A., Los Angeles)

———— 1976/77. Plant resources of the Suva Municipal Market, Fiji. *Ethnomedicine* 4(½):23–61

———— 1977a. The nature and importance of Tongan root crop production. pp. 83–89 in *Regional Meeting on the Production of Root Crops (24–29 November 1975, Suva, Fiji): Collected Papers*, ed. by Michel Lambert. Technical Paper No. 174. Noumea, New Caledonia: South Pacific Commission

———— 1977b. Urban root crop production in the southwest Pacific. pp. 73–82 in *Regional Meeting on the Production of Root Crops (24–29 November 1975, Suva, Fiji): Collected Papers*, ed. by Michel Lambert. Technical Paper No. 174. Noumea, New Caledonia: South Pacific Commission

———— 1977c. Urban gardening in Papua New Guinea and Fiji. pp. 146–168 in *The Melanesian Environment*, ed. by John H. Winslow. Canberra: Australian National University

————1978. Cooperative yam gardens: Adapting a traditional agricultural system to serve the needs of the developing Tongan market economy. pp. 116–128 in *The Adaptation of Traditional Agriculture: Socioeconomic Problems of Urbanization*, ed. by Ernest K. Fisk. Development Studies Centre Monograph 11. Canberra: Australian National University

———— 1979. Food scarcity, food dependency and nutritional deterioration in small Pacific island communities. pp. 191–197 in *Proceedings of the Tenth New Zealand Geography Congress and 49th ANZAAS Congress (Geographical Sciences), Auckland, January 1979*, ed. by W. Moran, P. Hosking and G. Atken. Conference Series No. 10. Auckland: New Zealand Geographical Society

———— 1982. The impact of tourism on agriculture in the Pacific Islands. pp. 130–156 in *The Impact of Tourism Development in the Pacific*, ed. by F. Rajotte. Environmental and Resource Studies Programme. Trent, Ontario: Trent University

Thaman, Randolph R. and Pamela M. Thomas. 1980. Tavioka in Fiji's future: Satan or savior? *Review* 4:42–44. Suva, Fiji: University of the South Pacific

———— 1982. The cassava invasion: The cultural, nutritional and ecological impact of cassava on Pacific island food systems, in *Proceedings of the Second Papua New Guinea Food Crops Conference, Goroka, 14–18 July 1980*, ed. by R. M. Bourke and V. Kesavan. Port Moresby: Department of Primary Industries

Thompson, Laura. 1940. *The Frontier.* New York: Octagon (reprinted 1972)

Tudge, Colin. 1977. *The Famine Business.* Harmondsworth: Penguin Books

Udo, Henk, John Foulds and A. Tauo. 1980. Comparison of cassava and maize in commercially formulated poultry diets for Western Samoa. *Alafua Agricultural Bulletin* 5(1):18–27. Western Samoa: University of the South Pacific

Ward, Ralph G. and Epeli Hau'ofa. 1979. The demographic and dietary contexts. pp. 23–44 in *South Pacific Agricultural Survey 1979*, ed. by Ralph G. Ward and A. Proctor. Manila: Asian Development Bank/Canberra: Australian National University

White, G. M. 1965. *Kioa: An Ellice Island Community in Fiji*. Department of Anthropology. Eugene, Ore.: University of Oregon

Williams, C. N. 1975. *The Agronomy of the Major Tropical Crops*. New York: Oxford University Press

Wright, A. C. S. and F. J. Van Westerndorp. 1965. *Soils and Agriculture of Niue Island*. Soil Bureau Bulletin 17. Wellington, New Zealand: Department of Scientific and Industrial Research

Wu Leung, Woot-Tsuen, Ritva R. Butram and Flora H. Chang. 1972. *Proximate Composition, Mineral and Vitamin Contents of East Asian Foods. Part 1: Composition Table for Use in East Asia*. Rome: FAO/Atlanta: Department of Health, Education and Welfare

Yen, Douglas E. 1979. Pacific production systems. pp. 67–100 in *Pacific Agricultural Survey 1979*, ed. by Ralph G. Ward and A. Proctor. Manila: Asian Development Bank/Canberra: Australian National University

Yuncker, Truman G. 1943. *The Flora of Niue Island*. Bulletin No. 178. Honolulu: B. P. Bishop Museum

APPENDIX I

Geographical importance of cassava in the Pacific Islands (7 = Dominant [most commonly cultivated] or Co-dominant food crop in the specified area; 6 = Dominant in some places in the specified area; 5 = Not dominant but an important staple crop in some areas; 4 = Minor staple in some areas and eaten occasionally; 3 = Present as a food crop but eaten infrequently; 2 = Present only for livestock feed or only rarely eaten as in times of famine; 1 = Absent; ? = Unsure) (Source of data is in parentheses and P = Personal observation of authors)

Country or Territory (Specific Region or Area)	Importance Score	Source	No. of Cultivars	Local Name(s)
Papua New Guinea				
(Lowlying Coastal Swampy)	3 or 4	(Powell 1977; Howlett 1973)	30 (Bourke, 1982)	tapioka
(Savanna and Grassland)	6	(Powell 1977; Bourke 1982; P)		
(Tropical Rainforest)	4 to 5?	(Clarke 1971; Bourke 1976, 1982; P)		
(Highland)	4			
(Urban)	7	(Thaman 1977b, 1977c)		
Solomon Islands				
(Guadalcanal Savannas)	6	(Lasaqa 1969; P)		kasava (pidgin)
(Densely Populated Large Islands; e.g., Malaita)	4 to 5	(Manner 1980)		bia, kai bia (kwara 'ae')
(Other Large Islands and Guadalcanal Weathercoast)	2 to 4	(Eele 1978a; P)		
(Small Outliers, e.g., Tikopia, Anuta)	7	(Hansell and Wall 1976; Yen 1979; Eele 1978a) (P)		
(Urban)	7	(P)		
New Caledonia				
(Dry Coastal)	6 or 7	(Barrau 1950; 1958; Guiart 1956; P)	at least 8 (Guiart 1956)	le manioc
(Moist Coastal and Alluvial Valley)	6 or 7	(P)		
(Urban)	7	(P)		

Vanuatu (New Hebrides)				
(Islands with Urban Focus)				
Espiritu Santo and Efate	6 or 7	(P)		maniok
Other Islands	6	(Bonnemaison 1974; Barrau 1958)		
(Urban)	6 or 7	(P)		
Fiji				
(Moist Windward Large Islands)	7	(P)	10 (Haynes and Sivan 1977)	tavioka
(Dry Grassland Large Islands)	7	(Bedford 1976; P)		
(Small Outer Islands)	7	(Thaman 1977b, 1977c)		
(Urban)	7	(Thaman 1975, 1976, 1977a; Hardaker 1971)		
Tonga				
(Tongatapu Group)	7	(Fa'anunu 1977; Maude 1973; Sevele 1973; P)	8 (Fa'anunu 1977)	manioke
(Ha'apai Group)	7	(Sevele 1973; P)		
(Vava'u Group)	6 or 7	(Thaman 1977b)		
(Urban)[a]	5 or 6	(P)		
Western Samoa				
(Savai'i)	1 or 2	(Uto, Foulds and Tauo 1980)	at least 4	maniota, manioka
(Upolu)	2	(P)		maniota, manioka
(Urban)[b]	3	(P)		
American Samoa				
(Tutuila)	4	(Malcolm 1954; Merrick 1977)		maniota, manioka
(Other Islands)	3	(?)		
(Urban)	4 to 5	(P)		
Niue	6 or 7	(Sykes 1970; Wright and van Westerndorp 1965; Mitchell 1977; P)	at least 7 (Yuncker 1943)	kapia, manioka

APPENDIX I (CONT)

Country or Territory (Specific Region or Area)	Importance Score	Source	No. of Cultivers	Local Name(s)
Cook Islands				
(Southern Cooks)[c]	5 to 7	(Michael 1977; Johnston 1967; P)		maniota, manieta aipi, pia, arrowroot
(Northern Cooks)[d]	1	(Michael 1977)		
(Urban)	4 or 5	(P)		
French Polynesia				
(Tahiti)	6 or 7	(Barrau 1970; P)		maniota
(Other Society Islands)	6 or 7	(Finney 1973; P)		manioka
(Other French Polynesia)	7	(Decker 1970)		manioka
(Tuamotus)	1			
(Wallis and Futuna)	6	(P)		
Tuvalu[e] *(Ellice Islands)*	3	(P)		tapioka
Kiribati (Gilbert Islands)	3	(P)		te tabioka
Nauru[f]	4	(P)		
American Micronesia				
Guam	6 or 7	(Stone 1970)	many (Stone 1970)	mendioka
Yap District	7	(Barrau 1961)	20–25 (Barrau 1961)	
Belau (Palau) District	7	(Templeton and Lambert 1977)		tapica
Ponape	6	Fischer and Fischer 1957)		kav tuga
Hawaii	5	(Neal 1965)	Several (Neal 1965)	manioka
Queensland[g]	3	(P)		cassava, manioca

[a] Most residents of Nuku'alofa, the capital of Tonga, have bush allotments where large crops such as cassava are normally grown.

[b] Cultivated and eaten occasionally by descendants of Solomon Island indentured laborers.

[c] According to Johnston, cassava is by far the most dominant and most preferred food on the island of Aitutaki in the Southern Cooks.

[d] The Northern Cooks are atolls where staple root crop production is limited to *Puraka (Cyrtosperma chamissonia)*.

[e] Tuvaluan settlers on the island of Kioa in the Fiji Group have adopted cassava.

[f] Kiribati, Tuvaluan and Chinese indentured workers in Nauru plant some cassava in their home gardens.

[g] Cultivated by some Pacific Island residents in Brisbane.

AFTERWORDS

CHAPTER 10

AFTERWORD:
THE TROPICS AND NUTRITION

ANTOINETTE B. BROWN
Department of Anthropology
Georgia State University
Atlanta Georgia

ABSTRACT

In the tropics of America, Africa and Oceania, food sources and food production systems can be different from those utilized in the temperate zone. Cultivation of tropical areas has typically depended on shifting agriculture in which impermanent fields are planted with crops for shorter periods of years than they are left fallow. Because shifting agriculture demands a large amount of land per person, population densities are characteristically low.

Root crops, including yams, taro and manioc are often of primary importance. These or other carbohydrate staples provide the bulk of the dietary calories. Domestic animals on the other hand are few and are economically peripheral. Tropical diets tend to have seasonal or chronic inadequacies, they depend on a starchy staple and they are monotonous. In these respects, tropical diets are similar to those of poor populations in the temperate as well as the tropical world.

While the tropical environment plays a role in the high incidence of disease and in poor soil quality that can depress agricultural yields, nevertheless human processes mediate the effects of the natural environment on the human diet. One of the most important of these human factors is technology. Support of agricultural technology, land reform, education and population control are all viable strategies to increase income and improve nutritional status in tropical countries. These innovations can best be introduced through an integrated development scheme in which increased productivity and higher incomes are incorporated into a total program which will also improve health and rural services such as potable water, electricity and mass transportation

J*

INTRODUCTION

For much of recorded history, and from what we can reconstruct of prehistory, the tropics have been extensively populated and extensively exploited. In the tropics of America, Africa and Oceania, food can be grown in ways quite different from those utilized in the temperate zone. Grain crops such as sorghum, rice and millet appear in place of wheat and barley. Root crops, including yams, taro, sweet potatoes and manioc are often of primary importance. Domestic animals on the other hand are few and economically peripheral (Netting 1977:58).

Cultivation in tropical areas has typically depended on shifting agriculture, a type of agriculture in which impermanent fields are planted with crops for shorter periods of years than they are left fallow (Conklin 1961). Shifting agriculture demands a large amount of land per person, because at any one time most of the arable land is being held as regenerating fallow. This means that population densities are characteristically low, seldom exceeding 150 per square miles (60 per square km) (Harris 1972; Netting 1977:62). Other estimates range from below 50 per square mile to as high as 400 persons per square mile (20 to 150 per square km). These latter figures are for modern peoples employing highly developed shifting cultivation regimes (Bennett 1976:126).

Not many years ago prolonged famines were predicted for large parts of the tropical world. Alarm about world food supplies has been a recurrent theme ever since Malthus. At least three waves of pessimism have swept the world since the end of World War II: in the late 1940s and early 1950s, in the mid-1960s and again in the early 1970s (Sanderson 1975). In each case, in concern was prompted by temporary shortages in the aftermath of war or a series of droughts in major grain growing areas. In each case predictions of continuous shortages were based mainly on high rates of population growth in the developing world. In the crisis of the early 1970s there was an additional emphasis placed on the consumption of grains by livestock (Sanderson 1975).

Food supply in relation to population is more satisfactory now than it was in the crisis years of the mid-1960s and early 1970s, and with the possible exception of Africa, there does not appear to be any decline in the *per capita* availability of food, although cyclical catastrophes continue to occur (Griffin 1979:1). Alarms about the possibility of famine are generally a response to war or natural disasters. While famines tend to be temporary, in many parts of the tropical world hunger and malnutrition are chronic.

The climatic conditions of the rainy tropics are propitious for agriculture and at first it is surprising that much of the tropical population suffers from food problems, either temporary shortages at the end of the agricultural cycle before the new harvests, or unbalanced diets that are deficient in protein or vitamins. The tropical diet is mainly vegetable. People living in the hot, wet countries display great ingenuity in making use of their vegetable food resources. They eat a variety of plants, including many noncultivated species which they gather. Food of animal origin plays a small part in tropical human diets, even where some animals are kept.

The chief food throughout is a carbohydrate: cereals, tubers and roots, bananas, breadfruit. This cultivated starchy staple—the "cultural superfood"—provides the bulk of the dietary calories. The papers in this collection describe four dietary staples: rice, manioc, corn and taro. The dietary staple is generally accompanied by boiled leaves, fat or oil, salt and seasoning. It may also be supplemented by a concentrated animal protein source: sometimes wild animals or fish, at other times domesticated animals, their eggs, or milk products.

In addition to their inadequacy and their dependence on a starchy staple, tropical diets are also recognizable by their monotony. There is little variation from meal to meal, from day to day, or from season to season. In this sense tropical diets are characteristic more of poverty and underdevelopment than they are of the environment, for inadequate monotonous diets which depend on a starchy staple for the bulk of their calories can be found among poor populations in every part of the temperate as well as the tropical world. In what sense then, can we speak of a tropical diet as distinct from a diet of poverty and underdevelopment? Is the tropical diet a consequence of tropical habitation, or is it the result of the interaction of human and natural forces? The intent of this commentary is to elucidate the nature of tropical nutrition by examining the relations between human communities and the tropical environment.

These relations are by no means simple. Each community perceives, utilizes and modifies the natural environment, not as an orderly collection of resources to be systematically developed, but in light of the techniques, customs and tastes which characterize its own particular civilization.

The relationship between human communities and their environment in influenced by the nature of the natural environment and by the nature of the human community—both biological and behavioral. Within the natural environment we will concern ourselves with climate, pathogens, soils, and animal and plant communities. Within the human community we will examine biological demands, agricultural technology and income.

THE NATURAL ENVIRONMENT

Climate

The tropics include lands that have no month with a mean temperature less than 18° C (34° F) and receive enough rain for agriculture to be possible without irrigation (Gourou 1980:1). By this definition the tropical world is warm enough for some crop to be harvested in every month of the year. The pattern of rainfall on the other hand exercises a restrictive influence on food availability. The true equatorial regime, with no marked dry season, permits annual plants to grow at all times but occurs over only a small part of the hot wet lands. The bulk of the area falls under the tropical regime with a marked distinction between rainy and dry seasons which limits the agricultural effort—at any rate for annual plants—to the rainy season and thus allows only one or two harvests a year (Gourou 1980:81).

The belt of equatorial rains has, as a rule, an advantage over regions with a tropical rainfall regime, since the vegetation, both spontaneous and cultivated, benefits from the continuous rains. In contrast, rainfall systems exhibiting a marked dry season are more irregular and may result in droughts or delays in the onset of the rainy season, thereby reducing crop productivity and leading to depletion of stored reserves. Inhabitants of the equatorial zone, who are more likely to grow and consume tubers as their main carbohydrate source, are much less affected by annual shortages (Gourou 1980:89).

As thus defined, the hot, wet regions have an area of 41 million square km (16 million square miles); 17 million (6.6 million square miles) in Africa; 14 million (5.4 million square miles) in the Americas, 8 million (3 million square miles) in Asia and the East Indies and 2.5 million (1 million square miles) in Melanesia, Australia and Oceania. This is a considerable fraction of the earth's land surface of 145 million square km (56 million square miles) and an even more remarkable proportion of its useful surface. If the world land area without direct human value is estimated as 45 million square km (17 million square miles), the hot wet land belt covers approximately 2/5 of the useful portions of the world's surface (Gourou 1980:2).

Most hot, wet lands are sparsely populated. In 1977 the 33 million square km (13 million square miles) of the tropics outside Asia had an average density of only 18 persons per square km (47 per square mile) (as against an average of 143 per square km (370 per square mile) in tropical Asia). The 33 million square km of the tropics outside Asia contain some 550 million people, 13% of the world's population on 28% of its exploitable land surface. These 550 million people belong to nations which are economically poor, badly provided with roads, railways and industries, and in which agricultural productivity is low in terms of crop yields and of human labor (Gourou 1980:5).

The hot, wet parts of Asia, with their dense populations and higher productivity, prove that there is no inevitable correlation between climate, low population density and poverty. The 8 million square km (3 million square miles) of tropical Asia contain 1000 million people or one fourth of the world's population on 8% of its exploitable area, with an average density of 143 per square km (Gourou 1980:5).

The sparsely peopled areas and the densely peopled areas of the tropical world have one thing in common besides climate: They are all underdeveloped. In the tropical world all aspects of underdevelopment are to be found: low per capita income, low energy consumption, widespread lack of education, poor roads and transportation services, weak communications and small-scale industry (Gourou 1980:5).

That the Hawaiian Islands, which in income ranks twentieth among the fifty United States, is also a tropical land, underlines the lack of geographical determinism in the tropical world. However, physical conditions are not without their influence on human lifestyle and diet. At the level of technical accomplishment in the exploitation of natural resources, poor health and the nature of tropical soils both help to explain the small numbers of people and the low standard of living in much of the tropical world. During the past two or three centuries, progress in the conquest of infectious diseases and in the maintenance of soil fertility has been realized in the temperate zones through research and experimentation. Such discoveries have only slowly been transferred to the tropical zone because of the tremendous diversity of the natural and social conditions existing there (Gourou 1980:6).

Tropical Flora and Fauna

The most highly visible aspect of the tropics is probably the vegetation. The green zone of the wet evergreen tropical forest, more commonly known as the tropical rain forest, stretched almost unbroken over the humid tropics for thousands of years. One of the world's most ancient ecosystems, the tropical rain forest is undergoing rapid change. Recent satellite photographs show that it is no longer a continuous belt but is fragmented and much reduced in area. In the past three decades huge expanses have been felled for timber or replaced by plantations of oil palms, bananas, rubber, cocoa and other crops. Still larger areas have been cleared for shifting cultivation. Sizable areas of rain forest still stand in Amazonia, Africa, Borneo, and New Guinea, but even in these regions the rain forest is retreating (Richards 1973).

The tropical rain forest is the most complex ecosystem on the earth. The most important of the producers in the system are woody plants ranging in size from the largest which commonly grow to more than 150 feet to miniature "treelets" no larger than shrubs but like trees in form. Also in this group are the woody-stemmed vines.

The crowns and foliage of the rain forest trees form several strata, or stories. The stratification cannot be readily observed from the ground, and it is rarely well defined. The presence of trees of many different heights subdivides the living space for animals and smaller plants. In the three or four horizontal layers that can be distinguished, environmental conditions vary widely (Richards 1973).

A major subdivision is between the canopy, which is exposed to almost full sunlight, and the undergrowth, which is much less brightly illuminated. In the canopy there is more air movement and a greater range of temperature and atmospheric humidity than there is in the undergrowth, which is one of the most constant of terrestrial environments (Richards 1973).

The great vertical range of environments in the rain forest is important to the organisms of which it is composed. Conditions for plant growth and function in the canopy are different from those in the undergrowth. For animals the variety of environment is equally important. In the various strata the available foods, the opportunities for concealment and the possible modes of locomotion are quite different.

The most distinctive characteristic of the tropical rain forest may be its great species richness; no other major ecological system has so many kinds of plants and animals. A two-hectare sample of lowland rain forest often contains more than 100 species of trees a foot or more in diameter. Large trees are only a part of the complement of plants. In the underbrush there are small trees and treelets, herbaceous plants, many vines and often large numbers of nonparasitic plants growing on tree trunks. Mosses, liverworts, algae, fungi and lichens are also present (Richards 1973).

The composition of the animal population is more difficult to characterize because few groups of animals have been adequately studied. For those groups of animals that are well known, the number of species has been found to be very large. In Panama a 300-mile square (800 km square) of rain forest harbors from 500 to 600 resident species of birds, more than four times as many species as found in the broad-leaved temperate forests of eastern North America. In the other animal classes, too, it is believed the number of species in the rain forest is large compared with the populations of other ecosystems (Richards 1973). The African forests are not as rich in species as those of South America and southeast Asia, but compared with temperate forests their complexity is still astounding (Owen 1973:6).

Just as there is a great variety of flora and fauna, there is also diversity from place to place within the tropics. The rain forests of the Old World and the New World are similar in general appearance and structure, yet they have almost no animal species and few plant species in common (Richards 1973).

Within a particular forest it is common to find that some species of plants and animals are dispersed unevenly with no apparent relation to differences of climate or other environmental factors. One result of the large number of species in the rain forest is that most are widely scattered and have a low population density. In a sample plot of a few acres most tree species are usually represented by a single individual; only a very few species contribute large numbers of specimens to the population. Sparse population is also characteristic of many animal groups (Richards 1973).

The environment of tropical Africa, like the other tropical zones, is diverse and there is a complex mosaic of land, some fertile and some barren, but each more or less suitable for some of the enormous variety of tropical plants. A cultivator who has grown up in an area where traditions have been passed from one generation to the next can accurately judge by looking at the natural vegetation the suitability of a piece of land for the cultivation of a specific crop. He (or she, since many African cultivators are women) will probably have acquired a vocabulary of hundreds of names for plants, trees and vegetation associations and will be fully aware of the significance of what might be called indicator plants when planning agricultural activities (Allan 1965).

The richness of floral species and their sparse distribution can be explained by the damage done by herbivores. The seeds, fruits and seedlings of tropical trees form the chief source of food for a horde of mammals, birds, insects and other animals, many of which seem to be host specific, that is, they feed exclusively on a single tree species or on a small group of species. It is common for the entire seed crop of a large tree to be destroyed by herbivores (Richards 1973).

A number of features of rain forest vegetation can probably be interpreted as defenses against insects and other herbivores. The degree to which plants have succeeded in this competition is perhaps indicated by the surprisingly low total mass of animal life in the rain forest compared with the mass of vegetation. The mass of the living plants in a hectare of Amazon rain forest is more than 900 metric tons (360 tons per acre) while the animals in the same area weigh about 0.2 ton (160 pounds per acre). Only about 7% of the animals, by weight, feed on living material such as leaves. The remainder eat living or dead wood, litter and other decaying matter (Fittkau and Klinge 1973). The low ratio of animal to plant life, when measured by weight, suggests that there is a shortage of edible plants in the rain forest. For human populations the rain forest is a poor place to live off the land. The scarcity of plant and animal food is one reason that jungle populations dependent on gathering and hunting wild food have had low population densities (Richards 1973).

Tropical Health

The indirect effects of climate on human nutrition are expressed in the diseases which are prevalent in the tropical world. Tropical populations are generally exposed to a large number of diseases. Malaria is still firmly rooted and there is a wide assortment of intestinal parasites—amoebae, hookworms, threadworms, and whip-worms. Schistosomiasis transmitted through contact with water affects many tropical populations. Filariasis is widespread. Yaws are often present, as well, as can be sleeping sickness and yellow fever. Worldwide diseases that are not confined to the tropics are also present: tuberculosis, leprosy, trachoma, venereal disease, as well as the diseases of childhood. Both birth rates and death rates are high, especially the child death rate up to 10 years of age (Gourou 1980:9).

Poor health is linked with the tropical climate. The steady high temperature, the humidity, the many water surfaces fed by rains, facilitate the continued existence of pathogenic complexes with which humans are closely associated (Gourou 1980:9).

Malaria is the most widespread of tropical diseases (Prothero 1965). Though it occurs in certain temperate regions, it is most common in the hot, wet regions of the world. Malaria weakens those whom it attacks. Episodes of fever increase nutrient requirements and make sustained work effort impossible. Hence agriculture does not receive as much attention as it might, and food supply is therefore affected at the same time as nutrient requirements are elevated. In this way a vicious cycle is formed. Weakened by insufficient nourishment, the human organism offers small resistance to secondary infection and cannot provide the effort required to produce an adequate food supply. Undoubtedly malaria is largely responsible for the poor health, high mortality and reduced work capacity of many tropical populations (Gourou 1980:11).

The program of malaria eradication launched in 1954 by the World Health Organization raised many hopes, but ten years later malaria still affected 57% of the 1,576 million people living in the tropical malaria zone. By 1977 the campaign had suffered a setback, as the number of new cases of malaria rose in South America, Africa and the Far East (Gourou 1980:12).

There are many reasons for this increase: the unintentional slackening of administrative control, a slow-down in the effort as a result of successes in the early years of the campaign, a large increase in the cost of petroleum-based DDT used for spraying, and resistance of the people in some areas. Lastly, new strains of anopheles mosquitoes have evolved, resistant to DDT, so that new insecticides must be developed. The classical antimalarial methods must continue to be employed: drainage, spraying, antimalarial drugs, while we await the discovery of new, more effective techniques of control.

Intestinal diseases and parasites also contribute to weakening tropical human populations, resulting in the loss of nutrients to the human body. Again the debilitating cycle—nutrient loss through malabsorption or parasitism—leads to reduced work effort, increased nutrient requirements, reduced resistance to infection and lower food production, which in turn lowers nutrient availability.

Sleeping sickness is restricted to sub-Saharan Africa; the microbe and the tsetse flies that carry it and which are indispensable to its development are only found within tropical Africa. Sleeping sickness physically weakens the persons it affects and frequently results in death. Presence of the tsetse fly also prevents the raising of cattle which could otherwise supplement the human diet (Gourou 1980:15).

Sleeping sickness, malaria and many other tropical diseases are characteristic of underpeopled and underadministered countries. The tsetse fly is associated with uncleared forest or woodland which provides the necessary conditions of temperature, humidity and shelter. It is not so much that the tsetse fly is responsible for the low population density as that the low density causes the abundance of the tsetse. It is the combination of geographic and demographic factors which operates to sustain the high level of disease infection and to reduce the level of human nutrition in tropical populations.

Soils

The predominantly acid nature of tropical soils reduces the number of cultivated plants to those adapted to tropical soil conditions which in turn affects the human diet. One might suspect that given equal soil fertility, tropical soils would be more productive than soils of the temperate zone. It is true that tropical agriculture benefits from higher temperatures, and in the equatorial zone, from year-long precipitation; thus crops can be harvested in every month of the year. On the other hand, crop growth is influenced by the amount of daylight, and in the tropics days are uniformly shorter than during the growing season of higher latitudes, while insolation is also greatly reduced by the rainy season cloud cover (Gourou 1980:22).

There is no predominant tropical soil type. A large-scale map would show a mosaic of soil types paralleling the diversity in other parts of the world. In general, however, soils in the rainy tropics are poorer in mineral matter that can be utilized by plants, such as lime, magnesium, potash, phosphates and nitrates. Tropical soils also tend to be poor in soluble elements, for the copious warm rains lead to a rapid and deep leaching of soluble materials (Gourou 1980:22).

The ease of leaching is in part due to the feeble adsorptive capacity of tropical soils. They cannot retain fertilizers because their clays have an unfavorable structure, because they have a tendency to accumulate inert lateritic elements and because they are poor in humus. Organic matter is quickly decomposed into soluble minerals which are easily lost to the soil through leaching. This process is exacerbated when forest cover is destroyed. The soil no longer receives the organic matter indispensable for making humus, and existing humus is decomposed more rapidly by an increase of soil temperature through direct exposure to the sun (Gourou 1980:26).

Agriculture in the tropics first developed in areas of high base status soils. These soils—now called Alfisols, Vertisols, Mollisols, and certain Entisols and Inceptisols (Soil Survey Staff 1973)—cover approximately 18% of the tropical land area. These soils have generally developed from alluvium, sediments or volcanic ash rich in calcium, magnesium and potassium. They present slight to no acidity problems and, therefore, require practically no investment in lime. Nitrogen is the most common limiting nutrient. Phosphorus deficiency, micronutrient disturbances and moderate salinity problems occur, but these can be corrected at low cost. High base status soils nearly always exhibit high native soil fertility and a relatively low cost of supplying additional nutrients (Sanchez and Buol 1975).

A larger group of soils in the tropics are of low base status and are highly leached. They are now classified mainly as Oxisols, Ultisols, some Inceptisols and sandy Entisols. They cover approximately 51% of the tropics (President's Science Advisory Committee 1967) in vast areas in the interior of South America and Central Africa and smaller areas of the hill country of Southeast Asia. These soils are commonly deficient in bases and often present aluminum toxicity problems. Deficiency of phosphorus is often difficult to correct because phosphorus fertilizers react with iron and aluminum oxides and are fixed in minimally soluble forms. Micronutrient and sulfur problems are common. Correcting these problems usually involves substantial investments in fertilizer and lime. On the positive side many of these low base status soils, especially the Oxisols, possess excellent physical properties which facilitate tillage and reduce erosion hazards. Similar soils, mainly the Ultisols, are found in the temperate region in such areas as southeastern United States and southeastern China where they support large populations (Sanchez and Buol 1975).

Strategies for Improved Land Management

The main factors limiting the development of tropical agriculture can be seen as low soil fertility, a limited transportation and marketing infrastructure and lack of apropriate soil management technology. The tropics are not unique in the percentage of soil exhibiting low natural fertility. What has been lacking is the science of soil management (Sanchez *et al* 1982). The first consideration is an awareness that shifting cultivation is an efficient soil management system, considering the resources farmers have at their disposal in sparsely populated forested areas. However, population pressures are reducing ratios of crop years to fallow years to the point where shifting cultivation is degenerating into a downward fertility spiral in many areas. The need for continuous cropping is acute and systematic research efforts are required.

A second consideration is that, instead of manipulating the soil to meet plant demands, the opposite strategy should be favored, that is, adapting plants to low base status soil conditions. Certain crop and pasture species are more tolerant than others to high levels for exchangeable aluminum, low levels of available phosphorus and other soil problems. Examples of the more tolerant species are upland rice, cassava, sweet potatoes, cowpeas and several grass and pasture legumes. Significant differences in tolerance to these limitations are also known to exist between varieties of rice, wheat, dry beans and soybeans. Tolerance to exchangeable aluminum and low available soil phosphorus can become a major component of breeding programs. The nutritive value of these varieties, however, should also be considered (Sanchez and Buol 1975).

Adaptation of crops to low base status soil cannot be viewed as a substitute for fertilizer applications. However it would significantly reduce the quantity of fertilizer and lime needed to obtain optimum yields. The question that should be considered is what are optimum yields. In the areas we are considering, land is relatively cheap while the cost of fertilizer and its transportation from the factory to the farm are expensive. Optimum yields should be considered those which optimize the use of the scarce resources—fertilizer and other energy-based inputs (Sanchez and Buol 1975).

A third consideration is devising a means for increasing the efficiency of the fertilizer and lime that are applied. The use of methods for evaluating soil fertility, including soil tests, plant analysis, correlation studies and interpretation, has substantially increased the efficiency of fertilization and liming in tropical America (Sanchez and Buol 1975). The concept of maximizing the efficiency of energy-related inputs in tropical areas is not necessarily limited to fertilizers. For example, research on land-clearing methods in the Amazon jungle indicates that the traditional slash-and-burn method with hand labor produced higher yields than does mechanized clearing with bulldozers. (Sanchez and Buol 1975).

Many problems concerning soil are site-specific and recommendations have to be compatible with practices at the local levels. Pronouncements concerning the effectiveness of soil-related practices have to be carefully evaluated according to soil properties. A single soil improvement formula is not available. A soil management strategy has recently been designed for the Ultisols of the Peruvian Amazon Basin (Sanchez *et al* 1982). Some 75% of the Amazon Basin is dominated by acid, infertile soils classified as Oxisols and Ultisols. They are deep, well-drained, red or yellowish soils with favorable physical properties, but they are very acid and deficient in plant nutrients.

About half of the Amazon Basin consists of well-drained landscapes with slopes of less than 8%. The other half has poor drainage or slope limitations. However, most of the constraints imposed by Amazon soils on agricultural development are chemical, rather than physical, in nature (Buol *et al* 1975). Phosphorus deficiency is found in 90% of the soils; the effects of this deficiency depend on which crops are grown. Only 16% of the Amazon soils are capable of fixing phosphorus into relatively insoluble forms. Aluminium toxicity, the main cause of poor plant growth in acid soils, affects about three fourths of the region. Low potassium reserves are also widespread. Poor drainage and flooding affect one fourth of the region. About 15% of the Amazon has effective cation exchange (CEC) values below 4 m Eq. per 100 g. A low CEC indicates that there are few negative charges in the soils capable of retaining nutrient cations such as calcium, magnesium and potassium. Consequently, even when large quantities of such nutrients are applied, they can be rapidly lost by leaching, particularly in well-drained soils (Sanchez *et al* 1982).

Field trials in this area of the Amazon Basin confirmed that methods involving slash-and-burn land clearing are superior to mechanical removal of vegetation. The ash produced by burning adds nutrients to the soil, while bulldozing often causes soil compacttion and displaces topsoil, depositing it outside the field. Also, with smaller clearings the transition to continuous cultivation can be made gradually. The most promising sequence of crop rotation so far has been that of three crops a year as rotations of upland rice, corn and soybeans, or upland rice, peanuts and soybeans. These rotations are adapted to the rainfall pattern and keep the ground covered most of the year. Continuous monoculture of the same crops did not produce sustained yields because of a buildup of pathogens (Sanchez *et al* 1982).

Fertilizers and lime were added to the soil according to recommendations based on soil analysis. During the second or third year, all yields began to decline rapidly due to a shorter than expected residual effect of the lime applied and the triggering of a magnesium deficiency induced by the potassium applications. After these problems were corrected, crop yields stabilized. The fertilizer levels did not differ substantially from those needed to grow corn, soybeans and peanuts in Ultisols of the southeastern United States. On an annual basis, the total amounts of fertilizer are higher in the Amazon than in the Southeast because three crops are grown instead of one or two. Soil properties improved with continuous cultivation which combined intensive management and appropriate fertilization (Sanchez et al 1982).

Continuous cultivation of annual food crops with appropriate fertilization is one of several options for sustained agriculture in the humid tropics. Relatively fertile alluvial soils that are not subject to flooding have a major food production potential. Technology is being developed for pasture production in sloping areas of Ultisols. The system would be based on the use of acid-tolerant grass and legume species. Many scientists believe that the natural vocation of the Amazon Basin is trees and that, ultimately, a tree canopy should replace crop or pasture canopies. Research is being initiated to combine crop production systems with promising tree species. The recommendation has been made that agricultural land should be integrated with forest reserves in a mosaic pattern to avoid the potentially undesirable effects of large areas of cleared land (Sanchez et al 1982). The intercropping of trees with other crops is not a new idea in the tropics. Fleuret and Fleuret, in this collection, report that in Usambara bananas are rarely grown in pure stands but are intercropped with one or more of the other basic foodstuffs. Bananas are also commonly planted in conjunction with coffee, one of the district's major cash crops, providing shade and mulch to the delicate coffee trees.

Drawing up a research strategy for the agricultural transformation of the acid soils of the tropics is difficult: nevertheless, most of the International Agricultural Research Centers (IARC), now banded together under the Consultive Group on International Agricultural Research (CGIAR) share a commitment to developing crops characterized by the ability to thrive in soils that are acidic, that contain toxic amounts of aluminum, or that are deficient in phosphorus and nitrogen (Plucknett and Smith 1982). Triticale, for example, is being tested in the Brazilian cerrado, a savanna region of some 200 million hectares (500 million acres) with little agricultural development due to highly leached soils containing large amounts of aluminum compounds. In Asia, the International Rice Research Institute (IRRI) is developing rice lines that thrive in acidic soils under rain-fed conditions.

Another priority at the IARCs is finding plants that thrive with relatively little fertilization. Some of the IARCs are focusing on the potential of biological nitrogen fixation to help boost crop yields (Plucknett and Smith 1982).

Agricultural research alone is not a panacea for the food problems of the tropical world. Social, economic and ecological issues must also be tackled, such as control of deforestation, population growth, crop losses after harvest and agricultural extension services.

THE HUMAN COMMUNITY

Agricultural Technology

It is clear that the tropical environment plays a role in the high incidence of disease and in the poor soil quality that can depress agricultural yields. Nevertheless it is also clear that human processes mediate the effects of the natural environment on the human diet. Perhaps one of the most important human factors mediating between the natural environment and the human diet is technology. Technology is a part of many facets of human existence: agriculture, transportation, communication, health care and manufacturing, to name a few.

Examining more closely the role of agricultural technology in food production and nutrition, one observes that over a large part of the tropics agriculture conforms to the pattern of shifting cultivation without the aid of irrigation. The forest trees are felled with axes, and when dry, the vegetation is burned. After one or more harvests the patch is left fallow and the forest is permitted to regenerate before it is once more felled and burned. This form of agriculture is appropriate on poor tropical soils, for the cultivated crops generally benefit from the fertility provided by the woodash. Other forms of agriculture are also possible in the rainy tropics, such as flood-irrigated rice cultivation, fruit tree plantations, etc. Shifting cultivation is not imposed by natural conditions, but corresponds to a certain level of agricultural technology as well as to social and economic factors. Some of these factors are uncertain land tenure, lack of capital for agricultural investment, absence of cash crops and low population density (cf. Boserup 1965).

Destruction of the forest cover by fire is easier than any other method, when one's only implements are the hoe and a machete or axe. The slash-and-burn method of forest clearing is quick and in a short time provides sufficient land which can be cultivated with hand tools for subsistence agriculture. While the soil is enriched with woodash from the fire, burning destroys per hectare tons of organic matter which would have been more profitable in the form of timber, firewood, wood pulp, leaf manure and products from the distillation of wood, had these alternatives been available. The tropical cultivator within the present technical and economic framework can do little otherwise. The unavoidable deficiency of alternatives in sparsely peopled countries prevents systematic exploitation of the forest.

Sowing generally begins after the first rains following the felling and burning of the forest. Several kinds of seeds and tubers may be planted in the same plot. In this way can be obtained both a main crop which supplies the bulk of the dietary calories in the form of carbohydrates, and subsidiary crops to accompany it and provide complementary nutrients. The variety of combinations is infinite. Owing to the mixture of cultigens, the swidden plot is often covered with a mantle of crop plants which, to some extent, protects the soil.

The plot must be periodically cleared of weeds. Weeding is time consuming and difficult in fields with mixed crops or on uneven terrain, so there is little point in extending the cultivated area. Instead, a new plot is preferentially cleared in another location. The harvest marks the end of the agricultural cycle. The harvest proceeds slowly, often utilizing a knife rather than a sickle. Food crops are carried from the field on the backs of men, women and children. Once harvested, the security of the crop is uncertain in many places throughout the tropical world. Rats, insects and microbes may consume as much as one fifth of the harvest.

Food preparation by traditional methods is a slow process; the hulling and pounding of grain to make flour for gruel or bread may be done in a mortar, a hand-powered grinder, or between two stones. Such primitive tools are found in all continents, even among people who are in contact with more advanced technology, but who cannot afford to invest in it. Husking millet or grinding corn for a family can easily take an hour each day. But every day the cook also has to fetch firewood which may take just as long, and get water which can require another hour as well. The preparation of a Mexican meal, consisting of corn tortillas (prepared from whole corn), refried beans (requiring several hours to boil) and chile sauce, can require up to five hours of labor by the cook.

Wet-rice cultivation is a distinctive type of tropical agriculture, and one of the most important, for although it occupies but a small part of the Far East, it supports much of the rural population. Rice is tolerant of a wide range of soils. It requires high temperatures for the growing season and at least 1778 mm (70 inches) of water during the growing season, but this does not greatly restrict its range (Grigg 1974:75).

Wet-rice is tolerant of a wide range of environmental conditions, and there are a multitude of varieties adapted for many different microenvironments. Unlike upland rice, which is grown much like other cereals, wet-rice has to be submerged beneath slowly moving water to an average height of 100–150 mm (4–6 inches) for three quarters of its growing period. This requirement restricts wet-rice cultivation primarily to flat lands near rivers. Here little cost is involved in levelling the fields and water is supplied simply by the rainfall of the monsoon, but more commonly this is supplemented by river floods. Irrigation from canals or wells may also be practiced. Wet-rice requires an impermeable sub-soil to prevent the water from draining too rapidly. Rice also yields best on the heavy soils deposited by rivers in their deltas and lower reaches. The microenvironment of the wet-rice delta helps to explain the ability of the wet-rice cultivator to produce constant crop yields from the same field year after year without the use of manures or rotations. The water-covered wet-rice plot is protected from soil erosion, and the high water table limits the leaching of plant nutrients. Both floods and irrigation water bring silt in suspension and other plant nutrients in solution renewing soil fertility each year. The water in the wet-rice plot also contains blue-green algae which promote the fixation of nitrogen (Grigg 1974:77).

Despite great variations in the way in which rice is cultivated, some generalizations are possible. Farms are small and fields often small and widely scattered. Only in more recently settled areas do farms average more than a couple hectares. Most wet-rice farms are operated by family labor alone. The rapid growth of the Asian population in this century, however, has created a class of landless laborers who are sometimes hired during critical periods of harvesting and sowing (Grigg 1974:79).

Compared with most farming systems, wet-rice cultivation is labor-intensive. At the beginning of the agricultural year dykes, bunds and canals have to be repaired. Attention is then turned to the preparation of the plot. The soil has to be reduced to a muddy consistency before the seed is sown. This may be done with the assistance of a plow drawn by water buffalo or zebu oxen, although in some areas a water buffalo is simply driven around the field. More recently the hand cultivator has been of growing importance (Grigg 1974:82).

Rice seed is not usually drilled and is thus either broadcast or transplanted. Transplanting is by far the most common method. About one tenth of the plot is set aside for the nurseries, which are carefully cultivated and manured. After four to five weeks the seed-lings are pulled and transplanted in the plot, a task usually undertaken by women. Water is drained from the plot some weeks before the harvest. The rice is gathered ear by ear with a knife or, alternatively, with a sickle (Grigg 1974:82).

The use of rotations is rare among rice cultivators, although in parts of East Asia green manuring is practiced. In the same areas efforts are made to fertilize the plot with canal mud, compost, rice stalks and manure. Elsewhere the plots receive little manure other than the droppings of draft animals. The use of chemical fertilizers is a recent phenomenon. The animals kept are used primarily for draft purposes; the majority of all livestock in China is so used. Meat and milk form a negligible part of the diet. In China pigs and poultry, which can exist on scraps and residues, are the main domesticated sources of meat. More recently, attempts have been made to raise catfish in the rice plots, but this is still not common.

Wet-rice farmers of the tropics are usually described as subsistence farmers, for about one half of the world's rice is consumed on the farms where it is produced (Grigg 1974:79). The paper by Anderson in this collection contrasts a traditional southeast Asian agriculture based on rice and treecrops practiced by the Malays with Chinese agricultural practice based on intensive vegetable and livestock raising. The Malay example is fairly typical in that the majority of rice is consumed by the producer. Rice forms the bulk of the diet, supplemented by fish and other crops.

Tropical agriculture other than wet-rice cultivation depends only slightly on animals for manure or for work. The swidden system makes plowing unnecessary. If the soil has to be turned over, this is done with the hoe. Animals are used only occasionally for transport. In short, the swidden system of agriculture is one which functions well without the help of working animals.

This agriculture of axe and fire, sometimes supplemented by hoe, is appropriate when there is plenty of land available or when the cultivators are using soils they have neither the desire nor the means to improve. The agricultural practices of the American tropics differ little from those of the Asiatic and African tropics, even though, before Columbus, the plants cultivated were different.

Tropical diet is limited in part by the variety of cultigens which can be grown under the climatic and soil conditions of the tropics. Plant varieties are also in large part limited to those which can be successfully grown by the swidden system of minimal soil preparation and crop cultivation, or in the specialized conditions of flood irrigated rice.

For five centuries a great mixing of plant species has been going on throughout the tropics. Of the basic foodstuffs the Americas have received few plants other than rice and bananas, but have contributed many. Africa has adopted many new species, especially if their cultivation did not involve a departure from traditional agricultural methods; thus maize could be grown like sorghum, and sweet potatoes and cassava like yams. Because of the mixing of cultigens which has occurred, the tropical diet appears even more homogeneous today than it did in the 15th century before the widespread dependence of Africa on cassava and maize from the New World.

Some tropical crops are much more efficient than others as providers of calories for human diets. Sugar cane yields, by a very large margin, more calories of human food per acre than any other crop grown in the tropics. Sugar cane production has increased rapidly in recent years. Sugar estates, whether privately or publicly owned, have appeared in countries where they never existed before (especially in Africa) and multiplied in countries where they already existed (especially in South America). After sugar cane, the oil palm is the next highest yielder of calories and gives more edible oil per acre than any other crop. The oil palm, therefore, appears to have a bright prospect as a main provider of edible oil. After oil palms come bananas, followed by the tropical root crops (Masefield 1977:20).

It will be noted that the highest-yielding crops are perennials, and this is natural because they are in the ground every day of the year to take advantage of the conditions for continuous growth which are provided by the tropics. Annuals suffer from an inevitable interval in the use of solar energy between harvesting one crop and sprouting of the next. This interval may be reduced somewhat by interplanting one crop with another before harvest. Even so, the leaf area attained by a succession of annuals cannot match that of perennial crops (Masefield 1977:21).

In addition to its high caloric yield per acre, cassava is, of all the tropical staple food crops, the easiest to grow and requires the least labor measured in edible food calories per hectare. Cassava produces more calories than white potatoes, yams, sweet potatoes, taro, rice, maize, sorghum or wheat. Thaman and Thomas, in their contribution to this volume, describe the spread of cassava as a consequence of its ease of cultivation. Despite its low labor requirement, Dufour, in another contribution, describes how time-consuming manioc production and processing can be. Where manioc is the single most important source of food energy and where females assume daily responsibility for manioc cultivation and processing, the greater part of their daily time and energy is allocated to manioc production.

The spread of cassava must be viewed with some disquiet, for cassava flour is particularly poor in protein. When cassava is substituted for a cereal or for yams or taro, the greater protein quality of these foods is lost; famine may have been avoided, but deficiency diseases multiply, especially among small children whose protein requirements exceed those of adults. Cassava leaves are a useful vegetable, rich in protein, and cassava-leaf sauce or relish is a nutritious complement to cassava porridge. Unfortunately, as Thaman and Thomas point out, cassava leaves are an important food only in West Africa, Indonesia, Malaysia and parts of Brazil.

Turning to annual crops, the cereals (with maize probably leading at average tropical yields) have a slight lead over the grain legumes (considered as producers of calories only) in yield of calories per acre of crop. It has long been observed by tropical agriculturalists that the denser a tropical population becomes, the more it tends to turn to rice as its staple food. This tendency has, in the last 20 years, been exemplified in Africa, where rice production has nearly doubled, and in South America, where it has more than doubled. The reason is not that rice necessarily yields more than other cereals, but that it is capable of a more sustained yield under continuous cultivation without the use of manures and fertilizers (Masefield 1977:21).

We have been thinking so far in terms of calories only, but these are not the sole dietary need. Another basic requirement in human diets is for protein with a proper combination of amino acids. While such a combination can be obtained from an intake of vegetable protein derived from a variety of plant sources, it can be more easily obtained if a proportion of the protein intake comes from animal sources. Considering vegetable protein first, pulses have a higher protein content than cereals, and both contain more than the root crops. For animal protein, the highest productivity is obtained from fertilized fishponds, which are capable in the tropics of yielding up to 450 kg of fish per hectare (400 lbs. per acre) per annum. For this class of animal, productivity per hectare is higher in parts of the tropics than anywhere in the temperate zone. Among domestic livestock, pigs are, in the tropics as elsewhere, the most intensive producers of food per unit of land used, and are followed in descending order by dairy animals, poultry for egg production and grazing animals for meat (Masefield 1977:22).

One of the areas of tropical productive technology most often overlooked is that of stock raising. Stock raising was hardly realizable in pre-Columbian America because there were no cattle, or in tropical Africa because of cattle diseases. Today, the failure of stock raising to contribute significantly to tropical diets is due to a variety of factors, environmental and socioeconomic.

Hot, wet lands are not eminently favorable to cattle raising, for the cattle are exposed to serious diseases. Moreover, the climate is unfavorable to the preservation and distribution of pastoral products: meat, milk, butter and cheese. Perhaps the most important factor is the poor nutritional value of tropical pastures which contain low concentrations of leguminous plant species.

Given that the natural conditions are marginal for livestock, tropical societies have tended to be almost wholly agricultural. In pre-Columbian America the virtual non-existence of stockraising is easily explained by the absence of domesticated faunae; only the llama was domesticated in the Andes. In the wet parts of tropical Africa, sleeping sickness ruled out oxen, horses and donkeys, but did not exclude pigs, sheep or goats. In fact, in the most equatorial parts of Africa, neither sheep nor goats are lacking. They are very useful for meals on ceremonial occasions, but they are seldom raised in any systematic manner and rarely milked. Southeast Asia, unaffected by trypanosomiasis, has raised water buffalo for use in rice cultivation. India has an abundant cattle population; the animals yield milk and are used for draft purposes and are effective proof that the existence or absence of stock raising results from a combination of several factors: disease vectors in the environment, the presence or absence of domesticated animal species and a tradition of stock raising.

In respect to meat, milk, manure and work, tropical cattle do not make the contribution that their numbers would lead one to expect. Under existing conditions of disease loads, heat, poor genetic stock, climatic uncertainties, inadequate or unsuitable fodder and poor husbandry, the growth of tropical animals is slow. There is little milk yield from the common varieties. But clearly, should the control or elimination of cattle diseases allow stock raising as an ancillary to agriculture, it would offer the advantage to agriculture of both manure and animal power and supplement the human diet with high-quality protein. Possibilities of economic progress are opened up by the improvement of animal husbandry.

As with agriculture, the weaknesses of tropical stock raising arise not so much from tropical conditions *per se*, but mainly from inadequate technical skills. Controlled burning and rotation of grazing areas can increase the carrying capacity of tropical pastures, as can clearing shrubs, mowing grasses and sowing leguminous forage. Proper herd management such as selective breeding, culling, inoculation and supplementary feeding can also increase herd productivity. More capital-intensive techniques such as cattle lots, importation of animals for genetic herd improvement and the breeding of hybrid varieties for disease resistance and higher milk yield could in time transform tropical stock raising into an important source of high-quality dietary protein.

Income

Ten years ago, malnutrition was often thought to reflect a shortage of protein and other nutrients. Most nutrition programs concentrated on providing high-protein food to children, usually in school. The emphasis today is different. There are now a number of agreements on several broad propositions related to undernutrition.

Serious and extensive nutritional deficiencies occur in virtually all developing countries, though they are worst in low-income countries. They are generally caused by a shortage of food and not by an imbalance between calories and protein, although this may occur, especially among young children.

Malnutrition is largely a reflection of poverty: People to not have enough income for food. Given the slow income growth that is likely for the poorest people in the forseeable future, large numbers will remain malnouurished for decades to come. Poor nutritional practices and the inequitable distribution of food within families also are causes of malnutrition. Dewey, in this collection, raises another cause of poor nutritional status, and that is the failure to provide adequate food from subsistence agriculture due to a shift in land allocation to commercial crops when the value of the commercial crop is inadequate to purchase enough food to substitute for the subsistence crop. Of course, another factor is the ownership of land itself. Many farmers do not own enough land to raise sufficient crops either to feed their family or to sell to purchase an adequate diet.

The most effective long-term policies are those that raise the incomes of the poor, and those that raise food production per person. Other relevant policies include food subsidies, nutrition education, food fortification or enrichment, and increasing emphasis on producing foods typically consumed by the poor (World Bank 1980:59).

There is some evidence that inadequate family incomes, ignorance of good nutritional practice and the inequitable distribution of food within families all contribute to malnutrition, but that low incomes are the central cause. Poor people spend the bulk of their income on food. The consumption of calories, usually derived from the cheapest sources of calories, changes almost proportionately with changes in income. As people earn higher incomes, they eat better and spend proportionately less on food (World Bank 1980:61).

A lack of money is frequently compounded by poor nutritional practices. Several common beliefs about nutrition, such as witholding food from children with fevers, have harmful effects and must be attributed to ignorance rather than poverty. Several studies have found that better-educated parents have better-nourished children: That this reflects more than the higher incomes of educated parents is suggested by the fact that mother's education is more predictive of child health than the father's (World Bank 1980:61).

Education, especially girls' education, may also help remedy one of the most serious and intractable nutritional problems: the way food is distributed within the family. In most developing countries adult women receive a lower proportion of their food requirements than do adult males; girls likewise are generally less well-fed than boys (World Bank 1980:62).

Faster growth in average incomes is essential to reducing absolute poverty and malnutrition, especially in low-income countries where half or more of the people may be poor. But growth alone is not enough. This is partly because rising population tends to swell the numbers in absolute poverty even where they are a diminishing share of the population.

Poverty and high fertility are mutually reinforcing. Reduction of the rapid population growth which exists in most of the tropical world is not an end in itself, nor does it for every country increase the potential growth of income per person. But in the circumstances prevailing in most of the tropical countries, rapid population growth impedes economic growth by reducing investment per person in physical capital and human skills. For individual families the number of children affects how much parents can invest in each one's health and education, and thus in their future earning power.

Land reform—the redistribution of land ownership in favor of the poor—has been tried in many countries as a means of increasing incomes of the poor. In some countries it has raised incomes of the rural poor considerably, though better access to credit and extension services for small farmers has proved an essential adjunct. In most developing tropical countries, there is scope for further land reform. Because small farms tend to apply more labor per hectare and to use land and capital at least as productively as large farms, land reform will usually increase agricultural output after a period of adjustments (World Bank 1980:41).

Support of agricultural technology, land reform, education and population control are all strategies to increase income and improve nutritional status in tropical countries. These innovations can best be introduced through an integrated development scheme in which increased productivity and higher incomes are incorporated into a total program which will also improve health and rural services such as potable water, electricity and transportation. Presented as a total development package, plans for increased productivity will not conflict with the self-defined priorities of rural people.

Increased productivity and the concurrent needs for increased agricultural labor cannot effectively coexist with poor health and the excessive labor required for traditional subsistence activities such as meal preparation. By improving both adult and child health, and by reducing the time and labor requirements for non-agricultural tasks, two purposes can be served: first, nutrient requirements are reduced, and second, time is freed for agricultural production.

The challenge for scientists is to solve the technological and socioeconomic problems which limit tropical agricultural development. The formulation of improved strategies for agricultural development for the tropics does not mean an uncritical and unmodified transfer of modern strategies developed in and for the temperate zone. Instead we will need to develop complete development packages which will concurrently assist people to adapt to the pressures of increased agricultural productivity within the constraints of a tropical ecosystem and the socioeconomic constraints imposed by poverty and underdevelopment.

Successful agricultural intensification can be seen in Wilson's discussion of a Southeast Asian village which became more self-sufficient in rice production through the sale of land by the Malaysian government on a credit basis. Another successful adaptation is that of the Peruvian colonists studied by Berlin who now mill their own rice for personal consumption rather than sell it to the mill and then buy it back later at a higher price. The same colonists, as well as a neighboring group of Indians, also substitute some domesticated animals for scarce wild animals to provide a concentrated source of protein in their diet.

Kunstadter's study of a village in Thailand demonstrates another successful intensification in which worker productivity increased through the change from swidden agriculture to irrigated rice cultivation. Worker indebtendness was reduced through the availability of wage labor and the low cost of credit.

Anthropological field investigation can provide anthropolgists with a basis for working with policy makers to improve the level of tropical nutrition. Much research remains to be done. While we have learned something about subsistence agriculture, more research is urgently needed on the cultivation of poor soils and on the subsistence crops on which many low-income farmers depend. In countries where climate and soil conditions vary widely from place to place, much more research is needed to determine the best farming methods for each place.

While industrial research has played a part in reducing poverty, small enterprises are disadvantaged, job creation is limited, and some of the goods bought by the poor are not improved or made cheaper as rapidly as those bought by the rich. More reseach undertaken in the tropical developing countries could reduce these biases.

We still need to know more about tropical human communities, about dietary decision making within the household and how the effects of macrolevel processes influence household agricultural and dietary choices. We also need to know more about the differential impacts of macrolevel processes within the community among age, sex and socioeconomic groups, and about these groups' differing responses to the opportunities presented by agricultural development.

Tropical agricultural development cannot solve all the problems of malnutrition by mere application of new technology. What are needed are social as well as technological inputs to increase worker and land productivity and to improve human health and knowledge. All these inputs require money and skills which are often in short supply in the tropical countries. Nevertheless these expenditures will have to be made to satisfy the nutrient demands of human populations. Anthropologists and nutritionists belong at every stage in the process of agricultural development and nutritional change—planning, implementation, monitoring and evaluation. They need not be counted among those who promise miracles nor among the prophets of doom, but among those scientists who can affect successful change.

REFERENCES

Allan, William. 1965. *The African Husbandman*. Edinburgh: Oliver and Boyd
Bennett, John W. 1976. *The Ecological Transition: Cultural Anthropology and Human Adaptation*. New York: Pergamon Press
Boserup, Ester. 1965. *The Conditions of Agricultural Growth*. Chicago: Aldine
Buol, Stanley W., Pedro A. Sanchez, Robert B. Cate and Michael A. Granger. 1975. Soil fertility capability classification. pp. 126–144 in *Soil Management in Tropical America*, ed. by Elmer Bornemisza and Alfredo Alvarado. Raleigh: North Carolina State University Press
Conklin, Harold C. 1961. Study of Shifting Cultivation. *Current Anthropology* 2:27–61
Fittkau, E. J. and H. Klinge. 1973. On biomass and trophic structure of the central Amazonian rain forest ecosystem. *Biotropica* 5:1–14
Gourou, Pierre. 1980. *The Tropical World*. London: Longman
Griffin, Keith B. 1979. *The Political Economy of Agrarian Change*. 2nd edition. London: Macmillan
Grigg, David B. 1974. *The Agricultural Systems of the World*. Cambridge: Cambridge University Press
Harris, David R. 1972. Swidden systems and settlement. in *Man, settlement, and Urbanism*, ed. by Peter J. Ucko, Ruth Tringham and Geoffrey W. Dimbleby. London: Duckworth
Masefield, Geoffry B. 1977. Food resources and production, pp. 59–66 in *Human Ecology in the Tropics*, ed. by James P. Garlick and Ronald W. J. Keay. New York: Halsted Press
Netting, Robert McC. 1977. *Cultural Ecology*. Menlo Park, Calif.: Cummings Publishing Co.
Owen, Dennis F. 1973. *Man in Tropical Africa*. London: Oxford University Press
Plucknett, Donald L. and Nigel J. H. Smith. 1982. Agricultural Research and Third World Food Production. *Science* 217:215–220
President's Science Advisory Committee. 1967. *The World Food Problem*. Washington, D.C.: U.S. Government Printing Office
Prothero, R. Mansell. 1965. *Migrants and Malaria*. London: Longman
Richards, Paul W. 1973. The Tropical Rain Forest. *Scientific American* 229(6):58–67 (December 1973)
Sanchez, Pedro A. and Stanley W. Buol. 1975. Soils of the Tropics and the World Food Crisis. *Science* 188:598–603
Sanchez, Pedro A., Dale E. Bandy, J. Hugo Villachica and John J. Nicholaides. 1982. Amazon Basin Soils: Management for Continuous Crop Production. *Science* 216:821–827

Sanderson, Fred H. 1975. The Great Food Fumble. *Science* 188:503–509
Soil Survey Staff. 1973. *Soil Taxonomy*. Washington, D.C.: Soil Conservation Service, USDA
World Bank. 1980. *World Development Report*. Washington, D.C.: The World Bank

ADDITIONAL READING

Epstein, T. Scarlett and David H. Penny (eds.). 1973. *Opportunity and Response: Case Studies in Economic Development*. Atlantic Highlands, N.J.: Humanities Press
Hirschman, Albert O. 1967. *Development Projects Observed*. Washington, D.C.: Brookings, Institution
____. 1978. *The Strategy of Economic Development*. New York: Norton
Lipton, Michael. 1977. *Why Poor People Stay Poor: Urban Bias in World Development*. Cambridge, Mass.: Harvard Univ. Press
Nelson, Michael. 1973. *The Development of Tropical Lands: Policy Issues in Latin America*. Baltimore, Maryland: Johns Hopkins University Press
Tendler, Judith. 1975. *Inside Foreign Aid*. Baltimore, Maryland: Johns Hopkins University Press

CHAPTER 11

FOOD CROPS IN THE TROPICS

KARL H. SCHWERIN
Department of Anthropology
University of New Mexico
Albuquerque, New Mexico 87131

ABSTRACT

The single staple carbohydrate of tropical food systems is supplemented by an impressive diversity of secondary food crops. There is a brief discussion of those secondary crops identified by contributing authors in this volume, which are about equally divided among supplementary sources of starchy carbohydrates and protein, fruits, and condiments and stimulants. Tropical staples are shown to include numerous root crops (manioc, taro, the xanthosomas, the yams, sweet potato), several grains (rice, corn, sorghum, the millets) and a few arboreal species (banana, coconut, breadfruit). Many are grown both for subsistence needs and as commercial crops. There follows a detailed discussion of the five staple carbohydrates analyzed in the several papers: manioc, rice, corn, taro and bananas. For each crop, discussion centers on its distinctive botanical, physiological and chemical characteristics, what is known of its origins and spread, cultivation practices and labor requirements, productivity, processing, world production, nutritional characteristics and current research.

INTRODUCTION

As one reads through the papers in this volume, two striking characteristics of tropical food systems become apparent. First, there is heavy reliance on a single staple carbohydrate. Second, these staples are supplemented by an impressive diversity of secondary food crops. Let me discuss these in reverse order.

While a few of the authors have concentrated on single staple crop (Dufour, Thaman & Thomas) and have made no attempt to deal with

TABLE I

Tropical food crops selected for discussion by contributing authors

English name	Scientific name	Anderson	Berlin	Brown	Dewey	Dufour	Fleuret & Fleuret	Kunstadter	Little	McCutcheon	Thaman & Thomas	Wilson
areca palm (see betel nut)												
banana (plantain)	Musa spp.[a]	X	X	X	X	X	X	X	X	X		
bean, kidney	Phaseolus vulgaris[b]		X	X	X		X			X		
betel nut (areca palm)	Areca catechu[a]	X								X		
breadfruit	Artocarpus altilis[b]			X	X				X	X		
cashew	Anacardium occidentale[b]	X			X				X			
cassava (see manioc)												
chile	Capsicum supp.[b]	X						X		X		
coca	Erythroxylum coca[b]					X						
coconut	Cocos nucifera[a]	X			X							X
cocoyam (see xanthosoma)												
coffee	Coffea arabica[b]	X			X		X	X				
coriander	Coriandrum sativum[b]		X	X	X			X				
corn (maize)	Zea mays[a]	X	X		X		X	X				X
guava	Psidium guajava[b]	X			X				X			
lemon	Citrus sp.[b]					X			X			
lemon grass	Cymbopogon citratus[a]	X								X		

Common name	Scientific name
maize (see corn)	
mango	Mangifera indica[b]
manioc (cassava)	Manihot esculenta[b]
mint	Mentha spp.[b]
oil palm	Elaeis guineensis[a]
onion	Allium cepa[a]
peanut	Arachis hypogaea[b]
pineapple	Ananas comosus[a]
plantain (see banana)	
potato, white	Solanum tuberosum[b]
rice	Oryza sativa[a]
sago	Metroxylon sagus[a]
soybean	Glycine max[b]
squash	Cucurbita moschata[b]
	Cucurbita pepo[b]
sweet potato	Ipomoea batatas[b]
sugar cane	Succharum officinarum[a]
tania (see xanthosoma)	
taro	Colocasia esculenta[a]
tea	Camellia sinensis[b]
wheat	Triticum spp.[a]
winged bean	Psophocarpus tetragonolobus[b]
xanthosoma	Xanthosoma spp.[a]
yam bean	Pachyrrhizus erosus[b]
yams	Dioscorea spp.[a]

[a] For more detailed information, see Purseglove (1972).
[b] For more detailed information, see Purseglove (1968).

other cultigens, it is noteworthy that most have made reference to a variety of secondary crops which are integral to the food/subsistence system. Every species mentioned in one or more papers as being cultivated locally is listed in Table I, with three exceptions. Anderson, Dewey and McCutcheon each provide a definitive listing of local crops (see tables in their respective papers), so in the interests of being concise, only those crops that were also named by at least one other author are included in Table I. Even so, more than 36 species are listed, only five of which are treated as staples. In most cases these secondary crops are dealt with in passing. Our authors have made no attempt to be exhaustive, thus we can expect that even greater diversity exists in the local food systems than is documented here. Anderson is the most effective in discussing how these secondary foods are incorporated into the subsistence system. It is clear that they are important, if not essential, as a source of variety as well as providing essential vitamins, minerals and proteins that are not available from the subsistence staple.

Many tropical populations seem to have made a satisfactory adaptation without significant intake of animal protein resources; at least few of the authors refer to the husbandry of domestic animals (Table II). It is probable that in most cases animal protein does constitute a regular part of the diet, but it is acquired from harvesting wild fish and game (undoubtedly the case with Dufour's Tatuyo in the Colombian Amazon and McCutcheon's island Palauans). This would also apply throughout most of the Pacific area treated by Thaman and Thomas. In Southeast Asia fish management commonly occurs in conjunction with wet rice farming. In other cases animal protein may be acquired through trade in live animals, dried or salted meat or fish, or canned meat, chicken or fish.

The secondary crops appear to be about equally divided among supplementary sources of starchy carbohydrate and protein (breadfruit, yams, beans), fruits, and condiments and stimulants (betel nut, coca, coffee, chile, coriander, onion). More detailed botanical and cultural data, as well as information on uses, and nutritional characteristics of these species may be found in such sources as Burkill (1935, 1966), León (1968), Martínez (1959). Purseglove (1968, 1972) and Sturtevant (1919, 1972).

The staple crops are referred to by several authors as "cultural super foods." Whether they are superior in some sense or other is beside the point here—they are central in the diet of their cultivators because they supply abundant edible starchy material which serves as

TABLE II

Domestic animals selected for discussion by contributing authors

English name	Scientific name	Anderson	Berlin	Brown	Dewey	Dufour	Fleuret & Fleuret	Kunstadter	Little	McCutcheon	Thaman & Thomas	Wilson
cattle	Bos taurus			X	X							
chicken	Gallus domesticus	X	X									
duck	Anas platyrhynchos	X	X									
goat	Capra hircus	X		X								
pig	Sus scrofa		X	X								
sheep	Ovis aries			X								
turkey	Meleagris gallopavo	X										
water buffalo	Bubalus bubalis	X		X								

the principal energy source. Such staple starchy food crops are limited in number, with a few especially attractive species widely cultivated around the world.

We in the temperate zone are accustomed to thinking of the basic staple in terms of grains (wheat, corn, barley, oats, rye)—only the white potato approaches them in its importance as a starchy staple (FAO 1978:110–111). In the tropics quite a different regimen obtains. Although rice and corn are important tropical grains, root crops also become significant (only in tropical Africa do sorghum and some of the millets achieve importance as staples). Manioc is second only to rice in tropical subsistence. McCutcheon's study also treats taro (*Colocasia esculenta*). Although not dealt with in these papers, several other tropical root crops are important as staples and/or major supplements in the diet. These include the New World aroids, *Xanthosoma* spp., the yam complex, *Dioscorea* spp. (with different species becoming locally important, cf. Coursey 1968) and the sweet potato, *Ipomoea batatas* (cf. Yen 1974). The true yams of the tropics belong to the monocotyledonous family Dioscoreaceae and are not to be confused with the dicotyledonous sweet potato (Convolvulaceae), certain varieties of which are colloquially called "yams" in the United States.

Only a few arboreal species serve as staple carbohydrate sources, and these are rarely widespread. The breadfruit (*Artocarpus altilis*) has been widely disseminated since the colonial perid but has rarely replaced established staples. The coconut (*Cocos nucifera*) is unique in providing complex fats and amino acids as well as starches (which may explain its limited use as a subsistence staple). Only bananas and plantains approximate the root crops in nutritional quality and their popularity as a food.

Natives of the temperate zone are rarely familiar with the full range of these tropical crops. No attempt will be made here to provide an exhaustive treatment, but as an aid to the general reader, a brief sketch of the five staple crops treated in these papers is presented. Attention will be directed toward distinctive botanical, physiological and chemical characteristics, what is known of the origins and spread of the crops, cultivation practices and labor requirements, productivity, processing, world production, nutritional characteristics and current research.

The actual contribution made to individual human nutrition by a given crop is subject to innumerable variables. Analysis of different races or varieties of the same species may show distinct nutritional

TABLE III

Nutritional content of tropical staples (g/100 g food)[a]

	Water	Protein	Fat	Carbohydrate	Fiber	Ash	Cal/100 g	KJ/100 g[b]
manioc	62	1–2	0.3	25–30		1.0	146	610
manioc leaves	80	4–7.5	0.4–1.3	6–9	1.0	0.7–2.0	55	230
yams (Dioscorea)	72	2.4	0.2	15–25	0.7	1.3	105	440
D. bulbifera	65	1.3	0.04	27–33	0.4	0.53		
D. cayennensis	83	1.0	0.05	15				
D. trifida		2.5	0.44	38				
taro	73	1.9	0.2	24	0.9	0.9	100	420
taro leaves	87	3.0	0.8	7	1.4	1.6	40	170
xanthosoma	73	2.5	0.3	22	1.3	1.0		
sweet potato	70	1.5	0.3	22	1.0		117	490
potato	78	2.0	0.1	19			82	343
rice	10–14	5.5–9.3	0.4–2.6	78–82	0.2–1.1	0.5–1.6	360–370	1505–1550
corn	13–20	9.0	5.0	77	1.6	1.2	340	1420
wheat	13	11.5	2.0	70	2.0	1.5	362	1515
banana	70	1.2	0.3	27	0.5	0.9	100	420
fresh coconut	51	3.7	31.0	14	6.3	0.7	328	1370

[a] Recommended daily allowances for adult males are 56 g protein and 2700 Cal (11,300 KJ) (National Research Council 1980:23)
[b] 1 Cal = 4.18 kilojoules

K

TABLE IV

Vitamin and mineral content of tropical staples (mg/100 g)

	Vit A I.U.	Thiamine	Riboflavin	Niacin	Vit C	Iron	Calcium	Phosphorus
Recommended Daily Allowance[b]	5000 i.u.	1.40 mg	1.60 mg	18.0 mg	60 mg	10 mg	800 mg	800 mg
manioc	tr.	0.06	0.03	0.6	36	0.7	33	60
manioc leaves	9000–13,000	0.16	0.30	1.5–2.0	200	2.0–3.0	206	
yams	tr.	0.10	0.03	0.5	9	0.8	22	22
taro	20	0.15	0.03	0.9	5	1.1	23	70
taro leaves	—	—	—	—	31	1.0	76	59
sweet potato	500	0.10	0.05	0.6	23	1.0	34	57
potato	tr.	0.10	0.03	1.4	10	0.7	8	56
rice	tr.	0.06–0.11	0.03	2.2–3.0	—	1.4–2.0	13–15	140–170
corn	400[a]	0.34	0.07	2.1	—	2.1	5	257
wheat	33	0.49	0.08	5.2	—	3.9	51	338
banana	1600	0.04	0.03	0.6	20	1.1	10	25
fresh coconut	—	0.03	0.01	0.5	3	1.9	13	95

[a] Yellow corn only, white varieties contain only a trace of vitamin A
[b] National Research Council 1980:following p. 185

characteristics. (For example, see the range of values given for rice in Table III & IV; also Table V, VII, and IX in Kunstadter's paper.) The quantity of various nutritional components may also be affected by climate, moisture, heat or cold and soil quality. Length of time in storage and techniques of processing or preparation may also affect nutritional quality. Some nutritional elements may be lost through deterioration in storage or driven off by heat or dissolved in the cooking water. At the same time other elements may be enhanced or added through such processes as fermentation. Finally, there is the question of what is actually absorbed by the human organism. This may be affected by the form of the nutritional elements and the health of the individual. The only truly definitive data would be derived through individual blood tests after consumption of a given food. However, complete testing requires a well-equipped laboratory with skilled technicians and would be prohibitively expensive for large scale testing.

The nutritional data presented here is offered as a basis for comparison among foodstuffs. It is intended as a guide for evaluating the relative contributions of the several staples being discussed. Perhaps this collection of papers will also stimulate more rigorous research into the real nutritional contributions of both staple and secondary tropical foodstuffs under various conditions.

A third striking feature of the staple crops emerges from the data summarized below. This is that there are two types of staple crop production: subsistence and commercial/export. A particular staple may constitute an important part of the subsistence diet in a given country, yet production for that country may be unimpressive when compared to world production statistics. Thus half the world production of bananas is in Africa where they represent a major subsistence staple, yet none of the major producing countries is in Africa—rather they are found in the tropical Americas. With the exception of Brazil, most of the latter countries produce for export rather than local consumption. Manioc and corn are likewise important subsistence crops in Africa, but the major producing countries are generally found in other areas. In much of the Pacific, as shown by Thaman and Thomas, manioc (cassava) has replaced the traditional subsistence staples, yet none of these island nations represent (or ever could represent) major producing countries. It is important, then, in assessing tropical food systems, not to forget the distinction between subsistence production and commercial agriculture.

MANIOC, CASSAVA, TAPIOCA (*Manihot esculenta* Crantz; syn. *M. utilissima, M. aipi, M. dulcis, M. palmata*)

This is the fourth most important crop in the tropics and its cultivation is spreading. Thaman and Thomas (this volume) have discussed its spread and current distribution in the Pacific and have included considerable information on its cultivation and nutritional characteristics. This discussion will supplement that treatment and place manioc in a worldwide context.

Manioc is a member of the dicotyledonous family Euphorbiaceae. It is a shrubby plant averaging about 2 meters (6 feet) at maturity. The woody and knobby stems generally branch several feet above the ground and may branch again, producing an open parasol shape. Leaves are palmate, generally with three to seven sections. Some of the roots enlarge to form starchy tubers.

Several hundred varieties are known. These divide into two con-variants: one with dark brown, rough skinned roots and brown, reddish or yellowish stems; the other with smooth light tan or pinkish tan roots and gray or silver gray stems (Rogers 1963). These morphological distinctions are unrelated to the amount of cyanogenic glucoside produced by the different varieties. These glucosides are enzymes which, when exposed to air, produce hydrocyanic acid (HCN, prussic acid). All parts of the plant produce these glucosides, but they are particularly concentrated in the root cortex and the leaves. Those varieties with a lesser production of glucosides in the roots and with the glucoside concentrated in the root cortex (10 to 75 ppm) are classified as "sweet manioc," while those with higher concentrations distributed throughout the root (100 up to 2000 ppm) are classified as "bitter manioc." While it is possible in one field or one region to identify precise associations between physical characters and level of cyanogenic glucosides, these associations are not consistent when tested in other areas. Furthermore, the distinction between bitter and sweet has often been based empirically on the methods used to process the tubers rather than on confirmation of glucoside levels. It is, therefore, impossible to classify bitter and sweet as distinct species (Schwerin 1970:25; CIAT 1982:211).

Manioc is an American plant, which was cultivated to its climatic limits throughout the New World tropics at the time of European contact. Rogers (1963, 1965) argues for a dual origin in southern Mexico-Central America and northeastern Brazil, while most botanists favor the latter area. Taking both botanical and cultural data into consideration, I favor the arid coastal regions or neighboring savannas of north-western South America for its place of origin. This is supported by archeological evidence suggesting that manioc was processed in this area at nearly 3000 B.C. (Schwerin 1970:26).

From South America the Portuguese carried manioc to Africa during the last half of the 16th century, but it did not spread widely until the 19th century. The Spanish seem to have taken it to the Philippines, from whence it spread gradually to Indonesia and Malaysia, reaching India by the end of the 18th century.

One of the great attractions of manioc is the wide range of conditions under which it will grow and the ease with which it can be cultivated. Short sections of the woody stem are stuck into the ground to propagate a new plant. For the first six to eight months, growth is concentrated in the above ground portions of the plant. After this it begins to store quantities of starch in large underground tubers which average 50 to 75 cm (20 to 30 in) in length and weigh 2 to 5 kg (4.5 to 11 lbs). Some varieties may be harvested after 6 to 10 months, but most take 10 to 18 months to mature. Generally the tubers can be left in the ground for two years or more without deteriorating or becoming woody. Manioc seems to do best in sandy loam with moderate moisture, but it will grow in a wide variety of soils. It is among the most drought resistant crops known, remaining verdant when even cotton is wilted. At the same time, it will survive temporary flooding, although some of the tubers may rot (Denevan & Schwerin 1978:26). Manioc does well in poor soils and is seldom fertilized but does exhaust available potash. In fact, a soil that is too rich will stimulate stem and leaf growth at the expense of tuber production.

Labor requirements are minimal. Planting is simple, a single weeding is usually sufficient and harvesting is accomplished by pulling the whole plant out of the ground. Research in Colombia has shown that a farmer may invest from 50 to 100 days of labor per hectare per year, but that there is no direct correlation between labor input and yield. The average subsistence farmer may harvest anywhere from 2.5 to 30 metric tons per hectare (1–13 short tons per acre) (world average for 1980 was 8.7 tons per hectare); under intensive cultivation it has yielded up to 65 metric tons per hectare (30 short tons per acre) (Table V) Manioc suffers from few diseases or insects, although mosaic virus has become widespread in recent years. Most grazing animals avoid the plant, although in some localities certian species may cause extensive damage (CIAT 1982:210, 217; Purseglove 1968:178).

Cooking breaks down the cyanogenic glucosides and volatilizes the

TABLE V

Productivity of tropical staples (1977)

	Average yield per hectare (metric tons)	Maximum yield per hectare (metric tons)	Average man-days per hectare
manioc	8.8	60	50–100
yams	100–200		
taro	7.5–30	37	
Xanthosoma	20	32.5	
sweet potato	9.6	50	
rice	2.6	6	100
corn	2.9	33.5	50
wheat	1.7	38.4	
bananas	30	50	
coconuts (copra)	0.9	2.5	

free prussic acid (HCN). This can be accomplished by simple roasting or boiling, but throughout South America the preferred method is to grate the tuber and then toast the mash to form large flat cakes (*casabe*, cassava) or a granular meal or flour (*farinha*). Sometimes the tubers are fermented before preparation of farinha (with a longer fermentation this is known in Africa as *gari*). The grated mash may also be boiled with water to provide a nourishing beverage. Cassava or farinha may be fermented with water to provide a thick beer. In West Africa boiled roots are pounded to produce a thick paste called *fufu*. Tapioca is derived from precipitated starch. Cassava and farinha may be stored for long periods if kept dry. In Malaysia and Indonesia, tubers are often sliced and dried for purposes of storage. The starch also has a number of industrial uses. In Brazil manioc is used as animal feed and manioc pellets are increasingly imported into Europe from Thailand for incorporation into livestock rations. Some authorities have discussed using manioc as a substrate for alcohol production (Cock 1982). In some areas manioc leaves are eaten after being boiled into a thick paste or sauce.

Ninety-nine percent of the crop is grown in tropical Africa, Asia and South America. In 1980 world production was 119 million metric tons on 13.8 million hectares (34.0 million acres) up from 54 million metric tons on 6.4 million hectares (15.8 million acres) in 1950 (Table VI). Principal producing countries are Brazil, Zaire, Indonesia,

TABLE VI

Cultivation and production of tropical staples

Crop	Area harvested—1000 hectares				Production—1000 metric tons			
	1950	1960	1969–71	1980	1950	1960	1969–71	1980
manioc	6391	7925	11020	13767	54250	73754	96378	119270
rice	103062	119975	134285	144529	163186	235165	311508	399112
corn	86642	107357	108487	128014	131044	216165	278504	395949
corn (excl. U.S.)	57344	78362	87738	98459	60834	116896	155855	227162
bananas					14299[a]	19685[a]	49058[b]	61849[b]
sweet potato			13941	11507	18178	24168	122270	148767
coconuts							27986	33955
wheat	171206	204127	215922	236873	167096	242582	329167	446107
potatoes	22760	25355	22383	17950	268726	283657	297239	230263

[a] bananas only
[b] bananas & plantains
(Sources FAO 1966, 1983)

Thailand and Nigeria, in that order (FAO 1983:128–129; FAO 1966:140).

About 80% of the whole manioc tuber is edible. This is composed of 62% water, 35% carbohydrate, although the actual assimilable starch is on the order of 25 to 30%. It is thus a very important carbohydrate and calorie source. Protein content is minimal, from 1–2%, and fat is at 0.3% (Table III). The tubers are, however, relatively rich in calcium and vitamin C and they contain reasonable amounts of B vitamins (more or less comparable to rice) (Table IV). Becaue of its low protein content, manioc cannot be relied upon as the sole dietary staple. It must be supplemented with other food, particularly protein sources.

Furthermore, where manioc is a major subsistence element, there is a tendency to develop sub-toxic levels of HCN in the human system. This can cause demyelination of the nerve sheath and result in degenerative neuropathy. The body is able to detoxify cyanide in the system, but in order to do this the liver needs a high input of methionine. Thus high quality animal proteins are necessary in the diet (Osuntokun 1973). These are generally available in most native American populations and probably also in southeast Asia. But in some places like West Africa and parts of Oceania, manioc has been adopted without the availability of the corresponding necessary porteins and the result has been a fairly common incidence of irreversible nerve degeneration among the population.

There is reason to believe that the widespread practices of fermentation in the processing of manioc may lead to conversion of some carbohydrates into proteins, but there has been little laboratory investigation of this question and almost nothing is known about the nature of such conversion. Manioc leaves are high in protein, with two to three times the dry weight provided by rice, but they are low in methionine and tryptophan so cannot substitute fully for animal proteins. (They are also an exceptional source of vitamin A.)

Manioc has long been neglected in agronomic and nutritional research, but active programs of investigation are currently being carried on in Brazil, India, Thailand, and especially at CIAT (Centro Internacional de Agricultura Tropical) in Cali, Colombia. The latter institution is attempting to acquire a complete collection of manioc cultivars, is working to improve productivity and cultivation practices and is exploring new ways of processing and utilizing the crop.

RICE (*Oryza sativa* L.)

Rice is the most important tropical crop, second only to wheat in quantity produced worldwide and the largest world producer of calories for direct human consumption (FAO 1978; Cock 1982). This grain is a member of the isolated tribe Oryzeae belonging to the monocotyledonous family Gramineae. It is an annual grass, 50–150 cm (20 to 60 in.) tall. It is mostly self-pollinated, which contributes to the maintenance of separate varieties, even within the same field.

Three basic sub-species have evolved, separated from each other by marked sterility barriers. Sub-species *indica* is confined to areas between 0° and 25° latitude, particularly in southeast Asia and the Philippines. It is adapted to areas with tropical monsoon climate and probably represents the most primitive group of rices. These varieties are characteristically tall, heavy, they tiller readily (produce side stalks), do not respond much to manuring and flower in response to short days. They are hardy, resistant to disease and tolerate unfavorable growing conditions. They will produce fair yields under conditions of limited management. The mature grains are usually long and slender and tend to resist overcooking, with the cooked grains remaining separate.

Sub-species *japonica* is mostly grown above 30° north and south latitude, in temperate climates with long summer days. It is less hardy, requires better cultural conditions, responds to heavy manuring, and generally gives much higher yields. Sub-species *javanica* is grown mainly in Indonesia, where it is known as *bulu* rice; it is adapted to equatorial climates. Some cultivars are intermediate between these three sub-species.

There are many thousands of rice cultivars; at least 5000–6000 are recognized in India alone and the International Rice Institute in the Phillippines made some 10,000 accessions of cultivated rice during its first three years. Cultivars are sometimes classified according to the conditions of moisture under which they are grown, but there are some which can be grown either in water or on dry land. Hill rice, unland rice or dry-land rice is grown as a rain fed crop where at least 750 mm (30 in) is available for at least three to four months. Swamp rice or lowland rice is grown on irrigated or flooded land and represents the bulk of world rice production. Floating rice is a group of cultivars that are grown in areas of deep flooding, usually from 2 to 5 m (6 to 15 feet). The plant grows rapidly to keep pace with rising water, but may take seven or more months to mature. The grain is harvested either after the flood has fallen or from boats. Floating rice is mostly limited to countries of southeast Asia, from Bangladesh to Vietnam.

Opinion varies as to whether rice was first domesticated in south India or Indochina. Archeological evidence for rice dates back 5000 years in India. It was definitely present in southeast Asia by 5300 B.C., and may have been in that area as early as 6000 B.C. (Solheim 1972). In both regions there are marshy areas, intervening mountains and periodic inundations which provide ideal conditions for cultivation of the crop. Nor is it certain whether the first domesticated rice was grown in dry upland areas or in areas of natural inundation. The wild ancestor is probably *Oryza perennis* which is widespread in both Asia and Africa and is probably ancestral to *O. glaberrima* also. (This species was independently domesticated in west Africa and is still grown in that region.) Early rice spread rapidly throughout China, being grown on the Yangtse river by the late Neolithic. Presumably these were at first *indica* races which evolved, perhaps through crossing with local races of *O. perennis*, into *japonica* forms. Rice reached Japan about 100 B.C. It was probably taken to Indonesia and the Philippines by 1500 B.C., but it was not carried into the Pacific (except for Guam), where even today it is unimportant. The Indonesian forms developed independently through crosses with local forms of *O. perennis*.

Sometime between 400 and 300 B.C. rice spread westward to Iran and reached the Mediterranean about the first century B.C. The Moors carried it along the southern Mediterranean to Spain. The Portuguese introduced rice into Brazil soon after their settlement of that territory in the 16th century, and the Spaniards brought it to Central America. Asiatic rice was also introduced into west Africa about this time (Purseglove 1972:164–166, 168).

Rice thrives over a wide range of climatic conditions from 53° N to 40° S and from sea level to 3000 m (10,000 ft.). The chief limiting factor is the water supply; soils are relatively unimportant. Long periods of sunshine are essential for high yields, particularly during the last 45 days in the field when the seed head is developing. Thus, where water is available, rice may be more productive during the dry season when skies are clear than during the wet season when the sun is obscured by clouds. (The longer days of higher latitudes also contribute to producing high yields). In the tropics, provided there is adequate water, two to three crops can be harvested in a year. Average temperature during the growing season varies from 20°–38° C (68°–100° F). Usually, however, rice is grown during the tropical monsoon. Land preparation is easier at this time, and water is plentiful. But drainage is often poor and deep flooding may occur during the middle and late stages of growth. Many tropical cultivars are adapted to nutrient poor soils and intense weed competition. Fertilization of these cultivars is counterproductive as it stimulates vegetative growth (Purseglove 1972:170–172).

Most swamp or lowland rice is grown in small holdings of 0.4–2.0 hectares (1–5 acres). The customary methods of cultivation are labor intensive, but these are usually performed by the farmer and his family. The world average manpower requirement is 740 manhours per hectare (300 manhours/acre), but on some small holdings in Asia it may reach 1000 manhours (400 manhours/acre). The field is prepared by building levees (bunds) and levelling the ground. It may be plowed or cultivated by hand. Rice is usually transplanted in a well-soaked field with little standing water, after which the depth of the water is increased to 15–30 cm (6–12 in) as the plants grow. Gently flowing water is preferable to stagnant water. Fields may be drained temporarily to facilitate weeding and fertilizing. Generally 1.5–2.0 m (5–7 ft) of water is required to produce a good crop. Many common weeds are controlled by the flooding and standing water, but aquatic and semi-aquatic weeds can be a problem. On the other hand, fertility in the paddy soils is enhanced through nitrogen fixation by *Azotobacter* and blue green algae. At the time of flowering, the water begins to be gradually reduced until the field is dry by harvest (Purseglove 1972:179–186).

Rice is commonly harvested by hand. For home consumption it is always stored in the husk, as it is less susceptible to deterioration in this state. It is then husked as required in small quantities to supply daily household needs. It is essential that it be well dried before storage, as this prevents heating and sweating; it also reduces attacks by storage pests and fungi. Moisture content at harvest is 18–25%, optimum moisture content for storage is 12.5% (Purseglove 1972: 187).

Rice yields are extremely variable from cultivar to cultivar, region to region and year to year, depending on the weather. The highest yields, 4000–5000 kg/ha (3500–4500 lbs/acre), of milled rice are obtained in subtropical and temperate zones where *japonica* rices are grown with long days and heavy fertilization. In southeast Asia, where *indica* rices are mainly grown, the average yields are 1000–1500 kg/ha (900–1400 lbs/acre). In 1979 the world average was 2650 kg/ha (2360 lbs/acre) (Table V) (Rutger & Brandon 1981).

The husks are removed from harvested grain by pounding with a pestle in a mortar, in a stone mill, or in mechanized mills. Rather than reduce the grain to powder as in wheat, the object is to keep the grain as whole as possible. Husked or hulled rice is brown rice. This is then milled to remove the outer layers, after with it is polished to produce white rice. Inevitably some of the grains are broken during husking and milling to produce broken rice. The milled grain gives about 20% husk, 50% whole rice, 16% broken rice and 14% bran and meal. During milling much of the protein, fat, minerals and vitamins are removed, so that flavor and nutrition are sacrificed for the white appearance. The thiamine that is lost, if not replaced from other components of the diet, may result in beriberi. White rice may be enriched by treating it with vitamins and minerals. It is then dried and sprayed with collodion to protect if from deterioration and depletion during washing preparatory to cooking. If the grain is parboiled before milling, it not only reduces breakage, but retains more nutrients and vitamins and results in better storage qualities.

Rice is usually cooked by boiling or steaming and eaten with other foods. Often it is the main source of calories and the principal food of millions of people.Ground rice, made from the broken grain, is used in confectionary, as is glutinous rice. The latter are certain cultivars that become sticky upon cooking, but they contain no gluten. Starch may also be extracted from broken rice. Rice powder provides a cosmetic. Beers, wines and spirits are manufactured from fermented rice. Rice husks are not much value as animal feed or fertilizer, but they can be used as fuel for rice mills or turned to various industrial uses. Rice bran, however, is valuable livestock and poultry feed. Rice straw can be fed to livestock, but it is inferior to other cereal straws; it may be used in various manufactures (Purseglove 1972:162–164).

In 1980 world rice production was 399 million metric tons 439 million short tons) grown on 145 million hectares (358 million acres). Average world yields rose 20% during the 1950s and another 19% during the 1960s, with 163 million metric tons produced on 100 million hectares in 1950, 235 million metric tons produced on 120 million hectares in 1960 and 312 million metric tons on 134 million hectares in 1969–71 (Table VI). Over 90% of the crop is produced in Asia, with the principal producers being China, India, Indonesia, Bangladesh, Japan, Thailand and Vietnam (FAO 1966:83; FAO 1983:110–111. Rutger & Brandon 1981:42).

The nutritional composition of rice depends on the method and degree of milling and polishing and whether or not it has been parboiled. This is also influenced by genetic and environmental factors, and the composition may be further altered through enrichment. The range of values in unenriched rice is, however, water 10–14%, protein 5.5–9.3%, fat 0.4–2.6%, carbohydrate 78–82%, with 360–370 Cal. per 100 g (Table III). The protein content of rice is lower than maize and wheat; rice is also lower in vitamin A, thiamine and phosphorus and intermediate in calcium content (Table IV) (Ecuador 1965:17; FAO 1970:40–41; Purseglove 1972:178).

Much attention has been paid to the improvement of rice, particularly with respect to increasing the yields and shortening the growing season. Research has been carried on in Japan and India for many years. The FAO started work in India in 1951 where experimentation involved crosses between *indica* and *japonica* cultivars. The International Rice Research Institute was organized in the Philippines in 1962 with support from the Ford and Rockefeller foundations. They have experimented with crosses among *japonica, indica* and *javanica* cultivars and have produced several very successful hybrids with yields of 6000–9000 kg/ha (5000–8000 lbs/acre). These have been the basis for the "Green Revolution" in Asia. There has also been some success in breeding for resistance to certain insect pests (Purseglove 1972:190–196).

CORN (Indian corn), MAIZE (*Zea mays* L.)

The third most important world crop, maize, is a more familiar plant because it is widely grown in both temperate and tropical regions, and because its growth habit is distinctive. It belongs to the Maydeae tribe of the monocotyledonous family Gramineae. Although the genus was long considered to consist of a single species, a second perennial species was recently discovered in western highland Mexico, and some authorities also classify teosinte as a species of *Zea*.

Corn is a giant grass usually two to three meters (6–10 ft) high, but ranging from 1–6 m (3–20 ft). Ordinarily a single upright stem is produced, but in some varieties (especially the more primitive forms) one to several secondary stems (tillers) may grow from the axils of the lower leaves. The male inflorescence or tassel is borne at the top of the stem, while the female inflorescence or ear is formed in the axil of one of the foliage leaves about half way up the stem. Several thousand cultivars are recognized, but no complete catalog has ever been compiled. This is made more difficult because maize is cross pollinated and exhibits exceptional heterozygosity. Under ordinary conditions of open pollination, it is doubtful that any two grains on the same ear have exactly the same genotype. Corn is usually classified into groups (flint, dent, flour, sweet, pop, pod and waxy) according to the ways in which the grain is used. Since these designations are based on single characters, and because individual plants in the field may be intermediate between two or more of these types, the classification has little botanical value. It continues to be widely used as an empirical classification, however, because of its utility to the farmer and household consumer.

Corn is another American plant which was cultivated to its climatic limits in both North and South America at the time of European contact. Mangelsdorf, MacNeish and Galinat (1964; MacNeish 1964) have shown conclusively that corn was domesticated in central Mexico by 3000 B. C. or even earlier. Complex patterns of hybridization between *Zea, Tripsacum* and *Euchlaena* introduced new characters and this combined with human selection contributed to the improvement of corn. It was long thought that Mexican maize was carried to Peru by 1000 B. C., but it is possible that a related wild form was independently domesticated there before the Mexican races reached the Andes to produce new hybrid crosses. Sine corn matures within a period of 90–120 days, it spread far into the temperate zones, being grown from southern Canada to Chile.

Columbus carried maize to Spain from whence it spread quickly throughout the Mediterranean. The Spaniards carried it westward to the Philippines, and it reached China by 1573. The Portuguese probably introduced maize into Africa. By the 17th and 18th centuries it had become an important cash crop in west Africa, being used to provision the slave factories and ships. When the slave trade was abolished, its importance as a cash crop declined. Although it was known in east Africa by the 17th century, it did not become a major crop until the present century. Today it is grown both as a staple foodstuff and a cash crop (Purseglove 1972:304–309).

In addition to its latitudinal range, corn can be grown from sea level to 3300 m (11,000 ft). It is a fast growing crop which will produce a harvest within a four-month growing season, provided there is adequate warmth and moisture. It does required a fairly loose, weed free soil, and thus necessitates a certain amount of soil preparation. Although corn agriculture is highly mechanized in the United States, most small scale tropical farmers cultivate it by hand. Simple tools are employed, and the grains are planted in individual hills. There is little use of manure or other fertilizers. Corn is, however, rather demanding of soil nutrients, so that where the same field is planted for more than one or two seasons, it is usually rotated with other crops, particularly legumes.

The ears are usually physiologically mature seven to eight weeks after flowering, but they may be left on the plant to dry out before harvesting. They may be harvested with the husks still attached. These are then turned back and the cobs tied in bundles and hung in a sheltered spot to dry further. The grain may be stored in this form, or the husked ears are stored on the cob, or the grain is shelled from the cob and stored loose (Purseglove 1972:320–321). Using steel tools the modern swidden farmer of Yucatan can feed his family with 50 days of work in the maize field each year.

Maize is the most productive of all grains. Yields vary tremendously according to the country and the conditions under which the crop is grown, ranging from less than 1000 kg/ha (900 lbs/acre) in much of west Africa, to more than 5000 kg/ha (4450 lb/acre) in the developed nations of North America and western Europe. The world average for 1977 was nearly 3000 kg/ha (2700 lb/acre) (Table V). Corn suffers from a variety of diseases including leaf blight, downy mildew, smut and a large variety of pests of which stem borers, earworms, cutworms and armyworms are the most serious. The black maize beetle cause damage in Africa. Corn is also ravaged by rodents, wild game, large birds and most domestic livestock.

There are hundreds of ways of preparing corn, from boiling or roasting the green ears to fermenting the grain to produce beer or *chicha*, or distilling the mash to obtain bourbon whiskey. In much of tropical America, the grain is ground into meal or flour, then prepared as cakes or buns (*tamales*, *arepas*, cornbread) which are steamed or baked. The meal may also be prepared as a mush, porridge or gruel (*atole*, grits); this is the commonest method of preparation in Africa. In the Americas popcorn is a favorite snack food.

The value of the plant hardly stops here, however, as it provides excellent feed for livestock, either green, dried or fermented into silage. Its commercial value is based on the hundreds of industrial products which can be made from it, or which depend on some element derived from it. These range from starches, syrup and sugars, or oil, to highly processed products such as alcohols, synthetic fibers, plastics and paper (Purseglove 1972:302).

In 1980 world production of corn was 396 million metric tons on

128 million hectares (436 million short tons on 316 million acres). This compares to 131 million metric tons on 87 million hectares in 1950, 216 million tons on 107 million hectares in 1960 (up 65%), and 279 million tons on 108 million hectares in 1969–71 (a 29% increase for the decade) (Table VI). Although the U. S. produces nearly half the world crop of corn, major tropical producers include Brazil (20 million tons in 1980), Mexico, India, the Philippines and Indonesia. Although no tropical African country stands out as a major producer, most produce more corn than millet and sorghum combined (FAO 1966:51; FAO 1983:114–115).

Maize is an important source of energy (340 Cal/100 g) and is a good supplement for protein and fats (Table III). It is, however, deficient in the essential amino acids lysine and tryptophan. It is comparable to wheat in its content of vitamins and minerals, with a reasonable proportion of B vitamins, but with somewhat less calcium, phosphorus and potassium. On the other hand, yellow maize is fairly rich in vitamin A (Table IV) (FAO 1970: 38–39; Ecuador 1965:18).

Corn has been the subject of intensive research in the developed countries, particularly the United States, with spectacular improvements in physical characteristics, adaptation to environmental conditions, period to maturity, yield, chemical composition, etc. (cf. Purseglove 1972:323–332). Most U. S. corn is today produced from hybrid seed with heavy fertilization and highly mechanized technology. In contrast most tropical agriculturists continue to depend on locally adapted heterozygous cultivars. Use of hybrid seen would in fact be impractical, since it requires highly organized large-scale operations such as government agencies or commercial seed farms to produce the seed and a well-developed infrastructure of transportation and commerce to make it profitable for the farmer. More attention needs to be directed to developing partially inbred lines which are adapted to specific environmental conditions and diseases in the tropics and which would increase yields without requiring large inputs of labor or fertilizer.

TARO, COCOYAM, DASHEEN, EDDOE (*Colocasia esculenta* (L.) Schott)

Taro is a less important crop of the tropics, although it has long been an important staple throughout most of the Pacific, and it is much appreciated in the Caribbean. It is, however, grown throughout the tropics as a supplementary crop.

This root crop is a member of the monocotyledonous family Araceae, from which several other species have been domesticated for their starchy corms, such as the *Alocasia* spp. of southeast Asia and the *Xanthosoma* spp. of the Americas. True taro is a polymorphic species which grows to a height of 1–2 m (3–6 ft). A starchy corm forms at the base of the plant just under or at the surface of the ground. The apex of the corn produces a whorl of large peltate leaves with long erect petioles. The numerous cultivars differ in color of tuber flesh, color of foliage, acridity of the tubers and leaves, etc. Color of the petioles may vary from green to red or black in different varieties. Flesh of the corms may vary from white to yellow to pink or lavender. The pinkish varieties are generally preferred, especially when low in mucilaginous substances. Most wild Colocasias and some cultivated varieties contain quantities of calcium oxalate raphides (needle-like crystals) which are both physically irritating to human flesh (especially the mucous membranes) and which are poisonus when ingested in sufficient quantity. This substance is eliminated through cooking and pouring off the water. Superior varieties are, however, free of this principle (Burkill 1935:640; León 1968:136–137).

Two quite distinct growth forms occur within taro—dasheen and eddoe; within these two forms there are numerous edible clones exceeding 1000 in number. Dasheen produces a large central starchy corm up to 30 cm (12 in) long and 15cm (6 in) in diameter, sometimes with a few small side tubers. It usually takes eight to ten months to mature. This is the taro of the Pacific. With eddoes, a large number of cormels or tubers are formed clustering around the small main central corm. The smaller cormels are planted whole to propagate the plant which matures in five to six months.

Techniques for removing the oxalates and selection of non-acrid and non-toxic cultivars were undoubtedly involved in the process of domestication which must have taken place in southeast Asia. Wild *Colocasia esculenta* still occurs in that region. It was probably domesticated along with the yam (*Dioscorea* spp.). Taro was taken in early times to China and Japan where the eddoes were developed and selected (*C. esculenta* var. *antiquorum*). It was widely grown in India, and from there it reached the Mediterranean about the time of Christ and has been extensively cultivated in the Nile delta ever since. It spread westward across the Mediterranean; at about the same time it seems to have spread from east Africa across the Sudan to the Guinea coast (Burkill 1935:639).

The ancestors of the Polynesians probably took taro with them when they left the Malesian region and traveled through the Sunda Islands and along the New Guinea coast finally reaching Samoa. From there it spread throughout the Pacific. In fairly recent times taro was introduced to the West Indies from the Orient.

In dry-land cultivation dasheen grows best as the first crop planted after clearing the tropical rain forest, but it requires a minimum rainfall of 2500 mm (100 in). In savanna areas or during a dry season it requires irrigation or growing in swampy places. In much of Polynesia it was characteristically grown in swampy areas (also see McCutcheon's description of Micronesian taro agriculture, this volume). Usually it takes eight to ten months to mature. It does not grow well on dry loose soil; wet heavy soils are preferred, with good fertility and adequate humus. Eddoes are hardier than dasheen; they can be grown with less rainfall and they withstand colder climates. They require five to six months of frost free weather to mature. They can be grown on lighter, poorer soils than dasheen, but do best on rich loamy soils with good drainage (Purseglove 1972:63).

Although I have no precise data on labor inputs, succesful taro cultivation requires a fairly intensive labor investment. (See McCutcheon's description of Palauan taro agriculture.) In Trinidad the crop is harvested by pulling it up and the small side tubers are left in the ground to produce a second crop (ratoon). The first crop yields 15–20 tons/ha (7–9 short tons/acre), and the first ratoon 10–15 tons (5–7 short tons/acre). In Melanesia crops grown without irrigation are said to yield 7.5 tons/ha (3 short tons/acre), but this doubles when irrigated. Yields of 37 tons/ha (16 short tons/acre) have been reported (Table V). In the Pacific region the crop suffers from taro leaf blight, taro leafhopper and the larvae of the gabi moth.

Corms of both dasheen and eddoes may be boiled, roasted or baked. In Hawaii *poi* is made by pounding boiled, peeled corms of dasheen, which are then fermented anaerobically in water to produce a sticky paste. In west Africa it may serve as a base for *fufu*. Dasheen does not store well, even when defoliated it cannot be kept more than a month. Eddoes in contrast can be stored up to several months, provided they have been dried. Young leaves, particularly of dasheen, are widely used as a vegetable or spinach. The leaves are harvested commercially in Trinidad. Young shoots are sometimes blanched and eaten like asparagus (Purseglove, 1972:62, 66, 69).

The nutritional composition of both dasheen and eddoes is about the same: 63–85% water, 13–29% carbohydrate, 1.4–3.0% protein (Table III). The B complex and vitamin C are present in appreciable quantities (although lower than in the grains) (Table IV). Dasheen leaves contain about 3% protein, 6% carbohydrate, are an excellent source of vitamin C and a good source of calcium.

No statistics are available on world taro production. Dasheen is grown in most tropical countries where there is adequate rainfall. It was of some importance in the forest regions of west Africa but is being replaced by *Xanthosoma*. In the West Indies it is grown as a subsistence crop. In the Pacific the demand is greater than local production and the excess of demand must be met by importation from outside the area. Eddoes on the other hand are grown in the Caribbean and the Orient primarily for local use.

Little or no scientific research has been conducted on taro. In a few regions there has been an attempt to differentiate the various cultivars, but there has been no effort to improve the hardiness of the crop or to increase its yield.

BANANA, PLANTAIN, GUINEO (*Musa* spp. & cvs.)

Bananas are the most important of tropical fruits. They are second only to grapes in tonnage produced worldwide but are rapidly approaching equivalency. Fleuret and Fleuret (this volume) discuss bananas in east Africa. This summary will place them in a more general world context.

Bananas and plantains are monocotyledons belonging to the family Musaceae. They are tree-like perennial herbs, growing from 2 to 9 m (6 to 30 ft) tall, and are the largest herbs know to science. The plant grows from a basal corm which may be 30 cm (12 in) in diameter when mature. Leaves sprout from the corm to form a pseudostem composed of layers of leaf sheaths. The leaves emerge at the upper end of the pseudostem to form a terminal crown. The flowering stem pushes up through the center of the pseudostem; it bears a large, pendulous spike that produces female flowers near the base and male flowers toward the tip. Most cultivars are, however, parthenocarpic and will produce fruit whether pollinated or not.

Empirically a distinction is usually made between those cultivars that are sugary and eaten raw when ripe—bananas—and those that are starchy and cooked before eating—plantains. Botanically and genetically this distinction cannot be supported. The systematics, genetics and variability of bananas and plantains are extremely complex and are best understood by reference to the genome complement[1] of the various taxa, whether wild or cultivated.

Wild *Musa acuminata*, which is widespread throughout southeast Asia, carries a diploid genome makeup designated AA (2n = 22). It bears small, fertile, seeded fruits with little edible flesh. There are some AA varieties, however, which are female sterile and produce edible fruit without pollination. These characters would have been attractive to humans and would have been maintained through vegetative reproduction. AA cultivars are occasionally planted throughout the tropics for their small sweet thin-skinned fruits but are agriculturally important only in New Guinea. Somehow triploid *M. acuminata* (AAA) were also developed. The triploids have larger fruits and are more vigorous, more productive and hardier than the diploids. Being more attractive to the farmer, they would be planted more commonly than the diploids. Most of the important bananas of commerce—Gros Michel, Cavendish, Lacatan, Robusta, Valery—are AAA cultivars (León 1968;108–109; Purseglove 1972:347–352).

Musa acuminata cultivars were taken by man to areas where the wild seeded diploid *M. balbisiana* (BB, 2n = 22) is native—India, the Philippines, New Guinea. New edible hybrids were formed (AAB, ABB). The center of origin of *M. acuminata* is in the humid tropics. *M. balbisiana* is more abundant in areas with a monsoon climate and a pronounced dry season, and it has conferred hardiness and resistance to drought in the hybrids, which are thus better adapted to marginal monsoon conditions than pure *M. acuminata*. In addition, the balbisiana genome provides greater variability and disease resistance and produces fruits that have more starch and acid, a higher dry-matter content, more vitamin C and different textures and flavors. The ripe fruits of these hybrids are more suitable for cooking and include most of the "plantains." Among the AAB clones are Mysore, Silk or Apple, Pome, *Pisang raja, Maia maoli*, Dominico, Harton. In the ABB group are the *Pisang awak* and the Bluggoe (*cuatrofilos, topocho*). The widely grown Ladyfinger or *Ney Poovan* is an AB hybrid. Only a few tetraploids are known, of AAAA or ABBB genomes only, and none have proven superior.[2]

Bananas were taken by early Polynesians to the Pacific sometime prior to the Christian era. They were carried from Indonesia to Madagascar sometime between the First and Fifth centuries A. D. They spread into the heart of Africa via the Zambezi river and the great rift valley, then across the Congo to west Africa. These early introductions seem to have been AAA clones, and most places where bananas are important in Africa still rely on these. They may have been introduced into Uganda about 1000 A. D. Hybrid AAB and ABB cultivars are found nearer the east African coast and were probably introduced more recently; they have nonetheless spread into the Congo Basin. Bananas were found in west Africa by the Portuguese in the 15th century and were carried across the Atlantic by them and the Spanish. Bananas were enthusiastically adopted by the American Indians and spread so rapidly throughout the New World tropics that many authorities assumed they were pre-Columbian cultigens (Purseglove 1972:349).

The banana is mostly grown between 30° N and S latitude. They do best in tropical humid lowlands where the mean monthly temperature is 27° C (81° F), temperatures less than 21°C (66° F) retard growth. They also require abundant sunlight and plenty of water—about 25 mm (1 in.) per week or an average annual rainfall of 1300 mm (52 in.). They are, however, often grown with less moisture and in areas with a pronounced dry season, as Uganda or eastern Venezuela. Bananas can be grown on a wide range of soils, provided there is good drainage and sufficient fertility and moisture. They do best on loams of volcanic or alluvial origin, and they respond well to fertilizing, particularly additions of nitrogen and potash. They are readily damaged by strong winds, which can blow the plants down (Purseglove 1972:355–356).

In newly cleared fields bananas can be planted without much soil preparation. In land that has been cultivated previously, the soil is ploughed or turned over by hand. Bananas are sometimes interplanted with other crops; they are commonly used as shade for cacao, then gradually thinned out over a period of four to six years. If there is a pronounced dry season, planting is at the beginning of the rains. Bananas are propagated by suckers produced from the base of the parent plant or by cutting the parent corm into several pieces. These are planted in holes about 45 cm (18 in) deep and wide. Because bananas establish an extensive fibrous root system just below the surface of the ground, they are not usually cultivated very much. Although perennial grasses can cause serious competition, most weeds are not a problem. As the plant grows, all but one or two suckers are cut off. Those that are left are allowed to develop so as to produce a later crop. The time to shooting of the flower is seven to nine months in the tropical lowlands, but may require 18 months at 1000 m (3300 ft.) or in the subtropics. The initial crop may be ready to harvest in 9 to 18 months, depending upon the cultivar, local climatic and cultural conditions etc. Time from shooting to harvest varies from 2½ to 4 months. If the crop is sold for shipping, it is harvested 80 to 90 days after shooting; if it is for local consumption, it is allowed to remain on the plant for a longer period of time. Because of the constant production of suckers, a banana planting can last a long time. The average life of a field varies greatly. Commercial fields may last from 3 to 20 years, but some subsistence fields have been maintained as long as 50 to 100 years (Purseglove 1972:362–367).

In Central America the average yield for bananas grown commercially is over 50 metric tons/hectare per year (22 short tons per acre); on poorer soil, plantations with good management still yield about 40 tons per hectare (18 short tons per acre). Yields are usually lower in the Caribbean, ranging from 12 to 35 tons per hectare (5–16 short tons/acre) (Table V), and this more or less parallels east African productivity as reported by Fleuret and Fleuret (this volume). Panama disease has devastated many of the commercial areas and has lead to the replacement of the Gros Michel clone with Cavendish cultivars, which are resistant. Both Gros Michel Cavendish clones are susceptible to leaf spot, but this is controlled by spraying with fungicides.

The ripe fruits of many cultivars are sugary and easily digestible. They are eaten raw as a dessert fruit or snack and are fed to the sick. Unripe fruits or the ripe fruits of starchy cultivars are cooked and provide a food that is similar to the potato. About half the bananas of the world are eaten raw and ripe, the other half are cooked before eating. The sugary bananas are one of the biggest single items in the international fruit trade. The greatest acreage of bananas is in Africa, where they are utilized as a starchy food. In east Africa especially they serve as the principal staple. They are also important in the Amazon lowlands and in parts of the Orinoco and upper Rio Negro basins they have achieved the status of joint staples along with manioc. In east Africa the unripe fruits are peeled, wrapped in banana leaves and steamed, then pounded into a thick mush. Daily consumption of bananas in Uganda is 4–4.5 kg (9–10 lb.) per head. In Venezuela chunks of banana are often boiled as part of a stew. In Uganda and Tanzania quantities of beer are made from bananas. The beer is important nutritionally, for it is rich in vitamin B due to the yeast content.

Ripe fruits may be roasted or fried. Banana figs are prepared by drying sliced ripe fruits, banana chips by drying slices of unripe fruits. These serve as famine reserves in parts of east Africa. Banana flour is prepared from unripe fruits and banana powder from ripe fruits. Male buds are eaten as a boiled vegetable in parts of southeast Asia. The huge green leaves are used as plates, wrapping material and umbrellas. In many areas fiber extracted from the pseudostem may be made into fabric or rough cordage.

About half the world production is in Africa, with the remainder about equally divided between Asia and the Americas. In 1980 world production of bananas and plantains was 62 million metric tons (68 million short tons). During the 1960's production increased dramatically, although it is impossible to give precise figures since data is lacking on the production of plantains prior to 1969 (Table VI). Four to five million tons enter international trade each year; three fourths of this is produced in Central and South America and the Caribbean. Brazil is the world's largest producer, but Ecuador is the leading exporter. Other large producers are found in Central America. The nations of tropical Africa are the principal producers of starchy plantains, but very little of this enters international trade (FAO 1966:296; FAO 1983: 190–191).

In the maturing fruit, carbohydrate is stored as starch, but this is largely converted to the sugars glucose and fructose on ripening. An unripe firm green banana as harvested for export contains 0.1–2.0% soluble sugars and 19.5–21.5% total starch. In fully ripe fruits this becomes 18.5–29.0% total sugars and only 1.0–1.5% starch. In the starchy types the starch is converted less rapidly to sugars, the acidity is greater and the water content less. The edible pulp of ripe bananas makes up about two thirds of the fruit. It is nutritionally similar to the potato but contains less water. The flesh of the banana is about 70% water, 27% carbohydrate, 1.2% protein and 0.3% fat, with a caloric value of 100 Cal/100 g (Table III). There is a good content of vitamin A, a fair amount of vitamin C, but bananas are poor in the B complex. Thus banana beer is important because it provides vitamin B which is otherwise lacking. All parts of the plant are conspicuously rich in potassium, but the content of other minerals is low (Table IV).

Because of the high incidence of female sterility and parthenocarpy, research and improvement of banana cultivars is extremely difficult and time consuming. Most of the research to date has been carried out in Trinidad and Jamaica. Only one or two seeds may be produced per bunch, and only 10% of the seeds produce viable tetraploid embryos. These take two years to fruit and produce suckers. The latter must then be tested for resistance to Panama disease. If they survive, they must then be assessed for fruit quality. This involves a minimum of 10 years for each success. Up to 1966 only one clone had reached this stage, and even that proved unsatisfactory for commercial cultivation.

Although the prospects are encouraging, it may be many years before significantly improved clones become available for release (Purseglove 1972:371–374).

NOTES

1. A genome is the haploid set of chromosomes, as occurs in the gamete. Thus in the bananas, the A genome contains 11 chromosomes, as does the B genome.
2. Another group of *Musa* species that occur in the western Pacific with a lesser chromosome number (2n = 20) have served as the progenitors of abacá or Manila hemp (*Musa textilis*), and the Fe'i bananas of Polynesia (especially important in Tahiti).

REFERENCES

Burkill, Isaac H. 1935. *Dictionary of the Economic Products of the Malay Peninsula.* 2 vols. London: Crown Agents for the Colonies. 1966. Reprinted. Kuala Lumpur: Ministry of Agriculture and Cooperatives

CIAT. 1982. *Cassava Program Annual Report 1981.* Cali, Colombia: Centro International de Agricultura Tropical

Cock, James H. 1982. Cassava: A Basic Energy Source in the Tropics. *Science* 218:755–762

Coursey, David G. 1968. *Yams: An account of the nature, origins, cultivation and utilization of the useful members of the Dioscoreaceae.* New York: Humanities Press

Denevan, William M. and Karl H. Schwerin. 1978. Adaptive strategies in Karinya subsistence, Venezuelan Llanos. *Antropológica* 50:3–91

Ecuador. Ministerio de Previsión Social y Sanidad. Instituto Nacional de Nutrición. 1965. *Tabla de composición de los alimentos ecuatorianos.* Quito: Ministerio de Previsiòn Social y Sanidad

FAO. 1966. *World Crop Statistics. Area, Production and Yield 1948–64.* Rome: FAO
____ 1970. *Amino-Acid Content of Food and Biological Data on Proteins.* Nutritional Studies No. 24. Rome: FAO
____ 1983. *1982 FAO Production yearbook,* vol. 36. Rome: FAO

León, Jorge. 1968. *Fundamentos botánicos de los cultivos tropoicales.* San José, Costa Rica: Instituto Interamericano de Ciencias Agrícolas

MacNeish, Richard S. 1964. Ancient Mesoamerican Civilization. *Science* 143:531–537

Mangelsdorf, Paul C.; Richard S. MacNeish and Walton C. Galinat. 1964. Domestication of Corn. *Science* 143:538–545

Martínez, Maximino. 1959. *Las plantas útiles de la flora mexicana.* México: Ediciones Botas

National Research Council. 1980. Recommended dietary allowances. 9th revised edition. Washington: National Academy of Sciences

Osuntokun, B. O. 1973. Ataxic Neuropathy Associated with High Cassava Diets in West Africa. pp. 127–138 in *Chronic Cassava Toxicity, Proceedings of an Interdiscplinary Workshop,* ed. by Barry Nestel & Reginald MacIntyre. Ottawa: International Development Research Centre

Purseglove, John W. 1968. *Tropical crops: Dicotyledons.* 2 vols. London: Longmans, Green and Co.
____ 1972. *Tropical crops: Monocotyledons.* 2 vols. London: Longmans Green and Co.

Rogers, David J. 1963. Studies of *Manihot esculenta* Crantz and Related Species. *Bulletin of the Torrey Botanical Club* 90:43–54

_____ 1965. Some Botanical and Ethnological Considerations of *Manihot esculenta*. *Economic Botany* 19:369–377

Rutger, J. Neil & D. Marlin Brandon. 1981. California Rice Culture. *Scientific American* 244(2):42–51 (February 1981)

Schwerin, Karl H. 1970. Apuntes sobre la yuca y sus orígenes. *Boletín Informativo de Antropología* 7:23–27. Caracas: Asociación Venezolana de Sociología Reprinted in *Tropical Root and Tuber Crops Newsletter* 3:4–12. Mayagüez, P. R.: USDA Federal Experiment Station

Solheim, William G., II. 1972. An Earlier Agricultural Revolution. *Scientific American* 226(4):34–41 (April 1972)

Sturtevant, Edward Lewis. 1919. *Sturtevant's Notes on Edible Plants,* (ed. by U. P. Hedrick) New York Department of Agriculture, 27th Annual Report, vol. 2, Part II. Geneva, New York: 1972. Reprinted. New York: Dover Publications

Yen, Douglas E. 1974. *The sweet potato and Oceania: An essay in ethnobotany*. Bernice P. Bishop Museum Bulletin 236. Honolulu: Bishop Museum Press

INDEX